Identification, analysis and control of discrete and continuous models of gene regulation networks

Von der Fakultät Konstruktions-, Produktions-, und Fahrzeugtechnik und dem Stuttgart Research Centre for Simulation Technology der Universität Stuttgart zur Erlangung der Würde eines Doktors der Ingenieurwissenschaften (Dr.-Ing.) genehmigte Abhandlung

Vorgelegt von

Christian Breindl

aus Neumarkt i.d.OPf.

Hauptberichter: Prof. Dr.-Ing. Frank Allgöwer
Mitberichter: Madalena Chaves, PhD
   Prof. Dr. rer. nat. habil. Alexander Bockmayr

Tag der mündlichen Prüfung: 7.12.2015

Institut für Systemtheorie und Regelungstechnik
Universität Stuttgart
2016

Bibliografische Information der Deutschen Nationalbibliothek

Die Deutsche Nationalbibliothek verzeichnet diese Publikation in der
Deutschen Nationalbibliografie; detaillierte bibliografische Daten sind
im Internet über http://dnb.d-nb.de abrufbar.

D 93

ISBN 978-3-8325-4283-2

Logos Verlag Berlin GmbH
Comeniushof, Gubener Str. 47,
10243 Berlin
Tel.: +49 (0)30 42 85 10 90
Fax: +49 (0)30 42 85 10 92
INTERNET: http://www.logos-verlag.de

Acknowledgements

This thesis is the result of a six year stay at the Institute for Systems Theory and Automatic Control at the University of Stuttgart. This phase in my life took a little longer than expected at the beginning, but I wouldn't have missed a single day of it. The IST is a great place to work and I had the most wonderful colleagues, not few of which have become dear friends.

First of all, I would like to thank Prof. Dr.-Ing. Frank Allgöwer for giving me the opportunity and the support to conduct this work. Prof. Allgöwer has created a working and research environment for PhD students that is unparalleled. The lectures given and the research performed at the institute meet world class standards and internationally renowned researchers frequently stop by to present their latest results and exchange their thoughts. Also outstanding is the culture of exchange and discussion that is maintained at the IST. All of this provided me with a solid basis to formulate and pursue my own research objectives. The extent of freedom I was given in this respect is certainly unique to this institute and something I appreciate very much, as it has enabled me to grow as independently and critically thinking scientist. Moreover, I was allowed to visit international conferences at interesting places all over the world to present my work, and I am thankful for the many opportunities to teach classes and supervise students by which I could gain a deeper understanding of the respective fields and improve my didactic and presenting skills.

Besides these work related aspects I enjoyed very much being at the IST because of the people. We had much fun during many coffee brakes, pizza and movie evenings, and during outings with the institute or with SimTech. There are a few people I want to thank in particular. Steffen Waldherr, Jan Hasenauer and Dirke Imig were wonderful roommates I could always turn to. Steffen gave me guidance at the beginning of this thesis, Jan has been and continues to be a much valued discussion partner whose advise I appreciate very much, and Dirke helped me to stay clam during the end game. I furthermore thank Andrei Kramer, Daniella Schittler, Madalena Chavez and Klaus Schmidt for the discussions, inspirations and joint projects who led to contributions in this thesis, and I thank Jan, Andrei, Daniella, Klaus and Madalena for proofreading it.

My research was supported by the German Research Foundation (DFG) within the Cluster of Excellence in Simulation Technology at the University of Stuttgart and the Helmholtz Society within the project "Control of Regulatory Networks with focus on non-coding RNA" (CoReNe). I thank Jean-Luc Gouzé for hosting me at his institute at INRIA in Sophia-Antipolis during a three months stay and I thank Prof. Alexander Bockmayr for carefully examining this thesis and joining my doctoral exam committee next to Prof. Frank Allgöwer and Madalena Chavez.

Last but not least I thank my family for their love and support.

Christian Breindl
Pfaffenhofen a.d.Ilm, May 2016

Contents

Index of notation

Acronyms

ODE	ordinary differential equation
DES	discrete event system
LP	linear program
MILP	mixed integer linear program
DNA	deoxyribonucleic acid
RNA	ribonucleic acid
mRNA	messenger RNA

General notation

Symbol	Description
$\mathbb{N}$	set of natural numbers
$\mathbb{N}_+$	set of positive natural numbers
$\mathbb{R}$	set of reals
$\mathbb{R}_+$	set of positive reals
N	set of natural numbers $\{1, \ldots, n\}$
$^k N$	set of all subsets of N with k elements
$H(x, y)$	Hamming disctence between Boolean vectors x and y
$\binom{n}{k}$	n choose k

Discrete models

Symbol	Description
B	Boolean or discrete update function
$\mathcal{O}$	set of observed transition pairs
w	vector containing the values $B(x)$ for all possible arguments x
d_B	maximum value of $B(x)$
$\mathcal{D}$	domain of state vector x
d_i	domain of state variable x_i
d_Π	size of $\mathcal{D}$
P	polynomial representing B
a	vector containing all coefficients of P
E	matrix mapping a to w
X	set of states
X^o	set of observable states
$\mathcal{X}_a$	set of attractors
$\mathcal{X}_0$	set of initial conditions
Π	control strategy

P, K, G	automaton representations of the Boolean model
$\mathcal{R}$	robustness measure

Models of ordinary differential equations

Symbol	Description
$f_i(x)$	vector field for node i
$r_i(x)$	production term for node i
γ_i	degradation rate for node i
ν	activation function
$\mathcal{N}$	set of all activation functions
μ	inhibition function
$\mathcal{M}$	set of all inhibition functions
φ	activation or inhibition function
$M_{i,j}$	maximal value for monotonic function $\varphi_{i,j}$
S_φ	class of function φ
$x_i^{\text{low}}, x_i^{\text{high}}, x_i^{\text{max}}$	critical concentrations for variable x_i
$\lambda^{\text{min}}, \lambda^{\text{low}}, \lambda^{\text{high}}$	critical values for activation and inhibition functions
$\mathcal{I}^{x_i,\text{l}}, \mathcal{I}^{x_i,\text{h}}$	low and high intervals for variable x_i
$\mathcal{X}, \mathcal{F}$	forward-invariant hyper-rectangular sets
$T_\mathcal{N}, T_\mathcal{M}$	tubes for activation and inhibition functions
$\tau^{\text{min}}, \tau^{\text{low}}, \tau^{\text{high}}, \tau^{\text{max}}$	tube parameters
$\mathcal{R}$	robustness measure

Abstract

In this thesis, various aspects concerning the identification, analysis and control of discrete and continuous models of gene regulation networks are considered. Inherent to all problems arising in these fields is the large degree of uncertainty in the knowledge about the biological processes. The goal is therefore the development of algorithmic solutions which can address these uncertainties adequately.

We first address the problem of identifying Boolean and discrete models of gene regulation networks. To this end, so-called update functions which describe the regulatory interactions between the individual species in the network need to be reconstructed from data. The search space for these update functions grows however exponentially with the system size. In order to allow for an efficient exploration of this search space, a polynomial representation of Boolean and discrete functions is developed. This representation is then employed to obtain a reformulation of the exploration problem as a linear program. While this already means a significant complexity reduction compared to existing solutions, the number of optimization variables still grows exponentially with the system size. Therefore, a more compact polynomial representation for the biologically meaningful class of unate Boolean functions is derived. This representation allows for a linear programming formulation with a much more favourable, linear problem scaling. The price is some loss of generality.

As multistability is a frequently observed phenomenon for gene regulation networks, the problems of model validation and discrimination with respect to a desired multi-stable behavior are studied next. To this end, a modeling framework based on ordinary differential equations is elaborated that can describe the uncertain model hypothesis in an adequate way. An algorithm is developed that can efficiently decide if the hypothetical model is, in principle, able to exhibit the desired multistable behavior. In order to be able to compare and evaluate several alternative models, a novel robustness measure based on biological robustness considerations is introduced. It is shown that it can be efficiently formulated as a convex optimization problem for important system classes.

Finally, the problem of controlling discrete models of gene regulation networks is addressed. Unlike most approaches in the literature, limitations on the observability of the state variables and input limitations are considered. We show that the problem can be transformed into a state attraction problem for discrete event systems, for which efficient solutions already exist.

In summary, the thesis either refines or extends approaches form the literature, develops completely new concepts, or transforms the problems into known ones in order to arrive at algorithmic solutions that can support modelers at different states of the modeling process. The benefits of all methods are demonstrated with several examples in the individual chapters.

Deutsche Kurzfassung

1. Einführung und Motivation

Viele wichtige zelluläre Prozesse werden durch Genregulationsnetzwerke gesteuert. Dabei bezeichnet man als Genregulationsnetzwerk das komplexe Zusammenspiel aus DNA Segmenten, den Genen, und den darin kodierten Proteinen oder anderen kleinen Molekülen. Ein wichtiges Ziel der Systembiologie ist es, das Verständnis der komplexen Mechanismen dieser Netzwerke zu fördern. Die mathematische Modellierung stellt dazu Formalismen bereit, um aus dem verfügbaren biologischen Wissen eine kohärente Beschreibung des Systems zu generieren. Im Idealfall können so entstandene Modelle nicht nur beobachtetes und bereits bekanntes Verhalten reproduzieren, sondern ermöglichen auch Vorhersagen darüber, wie das Netzwerk auf eine Veränderung der Bedingungen reagieren wird. Die Entwicklung solcher prädiktiver Modelle ist nicht nur von theoretischem Interesse. Auch in Disziplinen wie der Biologie, Biotechnologie und Medizin finden sie vielfältig Anwendung.

Die mathematische Modellierung biologischer Systeme ist im Vergleich zur Modellierung technischer Systeme jedoch deutlich erschwert. Während gerade für mechanische und elektrische Systeme die Gesetzmäßigkeiten, aus denen sich das dynamische Verhalten ergibt, meist genau bekannt sind, ist dies für biologische Systeme oft nicht der Fall. Auch ist es deutlich schwerer, Teilsysteme zum Zweck der Modellierung abzugrenzen und durch Messungen genauere Einblicke in die zellulären Prozesse zu erhalten. Um zum Beispiel den Zustand eines Genregulationsnetzwerks zu erfassen, kommen häufig Techniken wie die DNA-Chip-Technologie zum Einsatz. Damit lässt sich zwar eine große Zahl von mRNAs gleichzeitig beobachten, die genaue Quantifizierung der Messung und Umrechnung in Molekülzahlen oder Konzentrationen ist aber sehr schwierig. Deshalb sollten die Ergebnisse eher in qualitativer als in quantitativer Weise interpretiert werden. Auch die aus solchen Messungen ableitbaren Hypothesen über die Struktur des Netzwerks sind dann qualitativer Natur und betreffen zum Beispiel eher Vermutungen über die Existenz oder Nichtexistenz einer Wechselwirkung zwischen den Molekülen des Netzwerks als über deren genaue Ausprägung.

Wegen dieser Unsicherheiten haben sich Modellierungsansätze, die das Systemverhalten in eher qualitativer Weise beschreiben, als besonders nützlich erwiesen. Boolesche Netze sind der wohl prominenteste Vertreter dieser Ansätze. Statt genaue Konzentrationen oder Molekülzahlen abbilden zu wollen, wird hier nur unterschieden, ob eine Molekülart in niedriger oder in hoher Anzahl vorhanden ist. Auch für die Beschreibung der Interaktionen zwischen den einzelnen Netzwerkelementen verzichtet man hier auf mechanistische Details, und beschränkt sich auf die Abstraktion der molekularen Mechanismen durch Boolesche Funktionen. Trotz dieses hohen Abstraktionsgrades gelang es in zahlreichen Studien, prädiktive Boolesche Modelle der untersuchten Genregulationsnetzwerke zu entwickeln (Sánchez & Thieffry, 2001; Albert & Othmer, 2003; Calzone et al., 2010).

Wegen der oben genannten Erschwernisse ist die Entwicklung mathematischer Modelle von Genregulationsnetzwerken ein langer und schwieriger Prozess. Neben den mathematischen Grundlagen benötigen Modellierer auch eine gute Kenntnis der relevanten biologischen Literatur. Wegen der Komplexität der Genregulationsnetzwerke und der Vielfalt an biologischen Daten und Fakten ist es nicht verwunderlich, dass die meisten Modelle manuell gepflegt werden. Es ergibt sich daraus ein aufwändiger iterativer Prozess aus Experimenten, Analyse und gegebenenfalls Anpassung der Modelle.

Der Ausgangspunkt für diese Arbeit ist der Bedarf an mathematischen Konzepten und Werkzeugen, die Modellierer bei typischen Problemstellungen in diesem iterativen Modellbildungsprozess unterstützen können. Dabei konzentriert sich diese Arbeit auf Probleme im Zusammenhang mit der Identifikation, der Validierung, und dem Entwurf von Strategien zur gezielten Beeinflussung von qualitativen Modellen von Genregulationsnetzwerken. Wie später noch genauer erläutert wird, stoßen existierende Ansätze für diese Fragestellungen hierbei schnell an ihre Grenzen, wenn die Systeme zu groß werden. Oft werden auch die speziellen Anforderungen biologischer Prozesse nicht ausreichend berücksichtigt. Das Ziel dieser Arbeit ist daher die Erweiterung bestehender oder die Entwicklung neuer Ansätze, um Modellierern für die Identifikation, Validierung und Regelung auch großer Modelle hilfreiche Werkzeuge in die Hand zu geben. Dabei kommen insbesondere system- und regelungstechnische Konzepte und Ergebnisse aus der konvexen Optimierung zum Einsatz, um effiziente algorithmische Lösungen zu erhalten. Die drei Problemfelder, sowie Schwächen bestehender Ansätze und die daraus resultierenden Anforderungen für die Methodenentwicklung in dieser Arbeit werden im Folgenden vorgestellt.

Die erste Fragestellung betrifft die Identifikation Boolescher oder diskreter Modelle von Genregulationsnetzwerken und steht damit auch ganz am Anfang des Modellierungsprozesses. Dazu müssen sogenannte "Update-Funktionen", die die Interaktionen zwischen den einzelnen Molekülen beschreiben und die Dynamik des Modells bestimmen, aus den Daten rekonstruiert werden. Da der Suchraum für diese Update-Funktionen jedoch exponentiell mit der Systemgröße wächst, ist eine rein manuelle Suche unmöglich, und auch existierende Methoden stoßen schnell an ihre Grenzen. In dieser Arbeit soll daher ein Ansatz entwickelt werden, der ein effizientes Durchsuchen dieses Raums ermöglicht. Dabei soll auch eine Einschränkung des Suchraums auf biologisch sinnvolle Update-Funktionen möglich sein.

Die zweite Fragestellung betrifft die Modellvalidierung, wobei untersucht wird, ob ein hypothetisches Modell bisherige Messungen in ausreichender Genauigkeit reproduzieren kann. Da Multistabilität, also das Vorhandensein mehrerer Ruhelagen, ein wesentliches Merkmal vieler Genregulationsnetzwerke ist, beschränken wir uns auf die Modellvalidierung bezüglich dieser Eigenschaft. Für den Fall, dass mehrere hypothetische Modelle das gewünschte multistabile Verhalten reproduzieren können, stellt sich auch die Frage, welches aus biologischer Sicht das sinnvollste Modell ist. In der Literatur gibt es jedoch kaum Ansätze zur Verifikation von Multistabilität, die die großen Messunsicherheiten und das unsichere strukturelle Wissen über die Netzwerke adäquat berücksichtigen. Auch Konzepte, um verschiedene alternative Modelle vergleichen und bewerten zu können, finden sich kaum. Die Entwicklung effizienter Algorithmen zur Modellvalidierung und Diskriminierung bezüglich eines gewünschten multistabilen Verhaltens ist daher ein weiteres Ziel dieser Arbeit.

Prädiktive Modelle sind für viele Anwendungen nützlich, um Möglichkeiten aufzuzeigen, wie das Netzwerkverhalten durch externe Stimuli oder andere Manipulationen gezielt beeinflusst werden kann. Das dritte Problemfeld ist daher die Berechnung möglicher Beeinflussungsstrategien. Leider gibt es gerade für diskrete Modelle kaum theoretische Ansätze, die typische Einschränkungen an die Beobachtbarkeit der einzelnen Systemgrößen und die Anwendbarkeit von Stimuli ausreichend berücksichtigen. Das letzte Ziel dieser Arbeit ist daher die Entwicklung eines Formalismus zur Berechnung von Steuerungsstrategien, die auch eine ausreichende Berücksichtigung dieser Einschränkungen ermöglicht.

2. Hauptbeiträge und Gliederung der Arbeit

Im Hauptteil der Arbeit werden die oben beschriebenen Szenarien genauer betrachtet. Um effiziente algorithmische Lösungen für die beschriebenen Probleme zu finden, werden insbesondere Konzepte aus der System- und Regelungstheorie und der konvexen Optimierung verwendet. Die Struktur der Arbeit und die Hauptbeiträge der einzelnen Kapitel werden im Folgenden zusammengefasst.

Kapitel 2 - Grundlagen

In diesem Kapitel werden für die Arbeit benötigte Grundlagen zu Genregulationsnetzwerken eingeführt. Ebenso wird deren Modellierung durch Boolesche und diskrete Netzwerke sowie durch Systeme gewöhnlicher Differentialgleichungen vorgestellt. Schließlich werden einige zur Erfassung des Zustandes eines Genregulationsnetzwerks typische Messtechniken vorgestellt. Die Präsentation erhebt dabei keinen Anspruch auf Vollständigkeit, sondern dient dazu, den Schwerpunkt dieser Arbeit auf qualitative Aspekte der Modellierung und Analyse zu motivieren.

Kapitel 3 - Identifikation Boolescher und diskreter Modelle von Genregulationsnetzwerken

Dieses Kapitel behandelt das Problem der Identifikation Boolescher oder diskreter Modelle von Genregulationsnetzwerken. Wie oben erklärt, besteht die Herausforderung hier in der Entwicklung einer Methode zur effizienten Suche nach Booleschen und diskreten Funktionen. Dazu wurde zunächst eine polynomielle Repräsentation diskreter Funktionen entwickelt, die eine bekannte Darstellung Boolescher Funktionen verallgemeinert. Unter Verwendung dieser Darstellung wurde dann das Optimierungsproblem zur Bestimmung Boolescher oder diskreter Update-Funktionen aus den Daten zunächst in ein gemischt ganzzahliges lineares Programm umgewandelt. Das Hauptergebnis besagt dann, dass die Ganzzahligkeitseinschränkungen sogar weggelassen werden können, wenn das Problem mit einem Simplex Algorithmus gelöst wird. Dazu wurde gezeigt, dass alle relevanten optimalen Lösungen unter den Punkten zu finden sind, die das für die Optimierung zulässige Gebiet aufspannen. Weiterhin wurden biologisch sinnvolle Einschränkungen des Suchraums auf die Klasse der *unaten* und der *kanalisierenden* (Englisch: unate und canalyzing) Funktionen untersucht. Diese Einschränkungen wurden durch weitere lineare Nebenbedingungen an das

Optimierungsproblem realisiert. Das eben beschriebene Resultat bleibt dabei gültig. Im Vergleich zu bisherigen Ergebnissen aus der Literatur bedeutet diese Methodik die Verallgemeinerung des Booleschen auf den diskreten Fall, eine signifikante Komplexitätsreduktion und die Möglichkeit, den Suchraum biologisch sinnvoll einzuschränken.

Da die Anzahl der Koeffizienten der entwickelten Polynome zur Repräsentation diskreter Funktionen exponentiell mit der Anzahl der Argumente wächst, wurde eine alternative kompaktere Darstellung für unate Boolesche Funktionen untersucht. Es konnte dabei gezeigt werden, dass sich diese Klasse von Funktionen durch eine sogenannte Vorzeichen-Darstellung affiner Polynome darstellen lässt, deren Anzahl an Koeffizienten nur linear mit der Anzahl der Argumente wächst. Auch die Komplexität des entsprechenden Optimierungsproblems zur Bestimmung unater Boolescher Update-Funktionen aus den Messdaten wächst dann nur noch linear mit der Systemgröße. Weiterhin ermöglicht diese Vorzeichen-Repräsentation die Definition eines neuen Robustheitsmaßes, durch das nach biologisch plausiblen Update-Funktionen gesucht werden kann. Der Vorteil der in diesem Kapitel entwickelten Formulierungen wird mit Hilfe eines Booleschen und eines diskreten Beispiels gezeigt.

Kapitel 4 - Validierung und Robustheitsanalyse multistabilen Verhaltens

In diesem Kapitel wurden neue Konzepte und algorithmische Lösungen zur Modellvalidierung und -diskriminierung mit Hinblick auf ein erwünschtes multistabiles Verhalten entwickelt. Da, wie oben beschrieben, das biologische Wissen über das System, wie auch die Messdaten großen Unsicherheiten unterliegen, wurde zunächst eine auf gewöhnlichen Differentialgleichungen basierende Modellierungsmethode weiterentwickelt, die eine adäquate Systembeschreibung ermöglicht. Mit diesem Modellierungsansatz ist es ausreichend, für eine Modellhypothese nur bestimmte Monotonizitätseigenschaften der Funktionen der rechten Seite festzulegen. Ihre genaue Form oder Parametrierung kann aber offen bleiben. Ruhelagen werden durch invariante Gebiete im Zustandsraum charakterisiert. Dabei ist es ausreichend, das gewünschte multistabile Verhalten nur durch eine ungefähre Angabe der Lage dieser Gebiete zu spezifizieren. Das Validierungsproblem besteht dann darin, zu entscheiden, ob ein System mit den festgelegten Eigenschaften prinzipiell diese invarianten Gebiete generieren kann. Dazu wurde in der Arbeit ein Algorithmus entworfen, der diese Frage unter milden Einschränkungen an das Modell eindeutig beantworten kann. Dieser Algorithmus ist kombinatorischer Natur. Die Komplexität wächst, bei einer gegebenen maximalen Anzahl von Transkriptionsfaktoren je Gen, linear mit der Systemgröße.

Im zweiten Teil dieses Kapitels wird angenommen, dass die Lage der invarianten Gebiete auch quantitativ spezifiziert werden kann. Für diesen Fall wurde das Modellvalidierungsproblem als Lösbarkeitsproblem eines Systems von Ungleichungen formuliert. Für zwei Systemklassen konnte außerdem gezeigt werden, dass dieses Problem konvex und daher gut lösbar ist. Weiterhin wurde ein neues Robustheitsmaß definiert. Dieses Maß basiert auf biologischen Robustheitsüberlegungen und quantifiziert, wie gut ein Modell die invarianten Gebiete auch unter Störungen erhalten kann. Es ermöglicht damit auch den Vergleich und die Bewertung unterschiedlicher Modellhypothesen. Im Unterschied zu den meisten Robustheitsmaßen aus der Literatur wird hier aber kein voll parametriertes Modell benötigt, sondern die Interaktionsstruktur selbst, bestimmt durch die oben genannten Monotonizitätseigenschaften, wird bewertet. Es lässt weiterhin die folgende anschauliche Interpretation zu: es ist umso leichter, die Funktionen der rechten Seite so zu wählen, dass das gewünschte multistabile Verhalten erreicht wird, je größer das Robustheitsmaß ist. Die Berechnung wurde als Optimierungsproblem formuliert, und es konnte gezeigt werden, dass dieses Problem für zwei häufig vorkommende Systemklassen konvex ist.

Die in diesem Kapitel entwickelte Methodik wurde angewendet, um verschiedene Boolesche Modelle zur Beschreibung eines Zelldifferenzierungsschritts zu untersuchen. Dabei haben sich schwach verknüpfte Netzwerke als besonders robust erwiesen.

Kapitel 5 - Steuerungsentwurf für Boolesche Modelle von Genregulationsnetzwerken

Dieses Kapitel beschäftigt sich mit dem Entwurf von Stimulationsstrategien für Genregulationsnetzwerke. Betrachtet werden dazu Boolesche Modelle mit Steuereingängen. Das Ziel ist es, mögliche Eingangssequenzen für diese Steuereingänge zu berechnen, die garantieren, dass das System in endlicher Zeit von allen möglichen Anfangsbedingungen in gewünschte Endzustände überführt wird. Dabei wird davon ausgegangen, dass nur ein Teil der Zustände messbar ist, und dass die Eingänge nicht beliebig gesetzt werden können. Beispielsweise, und motiviert durch experimentelle Praxis, wurde hier gefordert, dass jeder Eingang nur maximal einmal geändert werden kann. Es wurde dann gezeigt, dass eine Lösung für diese Problemklasse bereits auf dem Gebiet der ereignisdiskreten Systeme existiert. Der Hauptbeitrag dieses Kapitels ist somit die Transformation des Booleschen Steuerungsproblems in ein Zustandsattraktionsproblem für ereignisdiskrete Systeme. Weiterhin wurde eine Beschreibung der Eingangsbeschränkungen durch zusätzliche Automaten entwickelt und in die Problemstellung integriert.

Die Methodik wurde auf ein Boolesches Modell aus der Literatur angewendet, dass einen Entscheidungsprozess über Überleben oder Sterben einer Stelle beschreibt. Es konnten dadurch neue Strategien gefunden werden, die in diesem Modell das Erreichen des apoptotischen Zelltods garantieren.

Kapitel 6 - Zusammenfassung

Die Hauptergebnisse und Beiträge dieser Arbeit werden zusammengefasst und diskutiert. Außerdem werden weiterführende Fragestellungen aufgezeigt.

Appendix

Um im Hauptteil der Arbeit eine möglichst kurze und prägnante Darstellung zu ermöglichen, sind einige Beweise und Beispiele zu den Kapiteln drei und vier in die Anhänge A und B verlagert. Diese Anhänge sind somit Bestandteil der Methodenentwicklung in dieser Arbeit. Lediglich Anhang C enthält keine eigenen Beiträge. Dort werden zwei aus der Literatur bekannte Lösungen für das Zustandsattraktionsproblem für ereignisdiskrete Systeme vorgestellt.

1. Introduction

1.1. Motivation

Gene regulation networks are at the core of almost all cellular processes. These networks are formed by DNA segments, the genes, which can interact with each other via proteins or other small molecules they code for. Such an interaction can affect the inhibition or promotion of the expression of that gene, leading to complex gene regulation networks. A better understanding of these networks is not only an interesting goal in its own right, but also relevant for many biological, biotechnological or medical applications. The field of systems biology aims to support this process by mathematical modeling and analysis in order to derive a coherent picture from the available pieces of information and experimental data. The goal is to develop mathematical models that can not only reproduce experimental observations, but also predict the system behavior under new conditions.

The most important source for the development of mathematical models of gene regulation networks are measurement data. State of the art measurement techniques such as DNA chips can monitor a large number of mRNA concentrations and allow for the generation of large amounts of data. However, these data are often noise corrupted and difficult to translate into concentrations or molecule numbers, and should rather be interpreted as relative than as absolute measurements. Thus, the conclusions drawn about the interactions between the genes are often uncertain, too. Therefore, modeling frameworks which put more emphasis on qualitative than on quantitative aspects are very popular in this field. Boolean networks are the most prominent representatives of this class of frameworks, and have been successfully applied in numerous studies (Albert & Othmer, 2003; Chaves *et al.*, 2009; Calzone *et al.*, 2010). Despite their high level of abstraction and simplicity, sufficiently fine-grained and predictive models have been obtained.

However, the development of mathematical models for gene regulation networks is a long and difficult process. It requires deep knowledge of the related biological literature to set up one or several alternative models. The models need then to be validated under varying experimental conditions and, if necessary, extended or modified. Given the complexity of the networks on the one hand, and the large amount of uncertainty about the biological knowledge on the other, it is not surprising that model development is mostly carried out manually in a trial and error fashion and requires many iterations.

Therefore, mathematical concepts, methods and tools, that can support modelers on the various stages of the modeling process, are urgently needed. This thesis will contribute such methods for three frequently occurring problems in the context of model identification, model validation and the development of strategies to influence the network behavior in a desired way. As we will detail below, existing solutions to these problems often reach their limits when the systems become too large, or do not respect the peculiar needs of biological systems. The goal is therefore the refinement and extension of existing approaches or the development of new concepts to remedy these shortcomings. The three problem scenarios from the above mentioned fields, the shortcomings of existing approaches and the goals for the method development in this thesis are presented next.

At first, the problem of identifying Boolean or discrete models of gene regulation networks from typical data is considered. To this end, so called update functions have to be reconstructed from the data. In this respect, the main difficulty is the exponential growth of the search space of update functions grows the system size, which is also the limiting factor for many existing approaches. We therefore aim to develop a formulation that allows to explore this search space more efficiently. Moreover, it should be possible to restrict this search to biologically meaningful subclasses of update functions.

The second scenario is concerned with the problems of model validation and discrimination. These problems become relevant at later stages in the modeling process, when there are already several hypothetical models available. Multistability on the other hand is a commonly observed behavior for gene regulation networks, playing an important role in processes such as cell fate decision or pattern formation. Therefore, we will focus on multistability as the key property for the validation and discrimination problems. Solutions to this validation problem that can adequately address the uncertainty about the biological knowledge and the measurements are however largely missing. There are also hardly any concepts to compare and discriminate different alternative models, which are able to show the desired multistability pattern. We therefore aim to develop an algorithmic solution that can quickly decide if a hypothetical model is, at least in principle, able to reproduce the observed steady states. Moreover, we also require a novel concept to compare the different model hypotheses with respect of their biological plausibility. Thereby, especially the qualitative nature of the available information about the system needs to be addressed adequately.

While the above two scenarios are concerned with modeling and analysis aspects, the third scenario goes one step further and addresses the problem of influencing gene regulation networks by external stimuli in order to achieve a desired behavior. Especially for the Boolean or discrete setup, methods that can deal with typical input and observability limitations are largely missing. We therefore require a method that can deal with the typical case, that only a few states are measurable. It should furthermore be possible to respect the typical limitation, that inputs can not be applied arbitrarily.

In summary, we can see a need for efficient algorithmic tools for model development, validation and control design. This need will be addressed in the main part of the thesis, using concepts from systems and control theory and convex optimization.

1.2. Contributions and outline of the thesis

This section presents the outline of this thesis. Furthermore, the main results and contributions of the individual chapters are summarized.

Chapter 2 - Background

This chapter introduces some fundamentals from biology and mathematics needed throughout the thesis. The most important biological mechanisms of gene regulation as well as a few commonly used measurement techniques are explained. Concerning mathematical models of gene regulation networks, the discrete or Boolean framework, and a formalism using differential equations are presented. We motivate and justify the focus on qualitative aspects of gene regulation network modeling and analysis, which underlies all method development in this thesis.

Chapter 3 - Identification of Boolean or discrete models of gene regulation networks

This chapter deals with the problem of identifying Boolean and discrete models of gene regulation networks. For such a model, update functions describing the interactions between the species of the network have to be reconstructed from typical data. The goal of this chapter is the development of a method, which can efficiently search the space of update functions. Furthermore, it should be possible to restrict this search space to some biologically relevant subclasses. The individual contributions are as follows.

- A polynomial representation of discrete functions is developed.

- Using this polynomial representation, the problem of finding optimal update functions is cast as a linear optimization problem. We show that the optimal solutions of interest can be found efficiently using simplex solvers. Compared to existing results, this means a significant complexity reduction and the generalization of the Boolean to the discrete case.

- Restrictions of the search space of update functions to the biologically relevant classes of unate and canalizing functions are formulated as additional constraints to the optimization problem. We show that the optimal solutions of interest can still be found using simplex solvers.

In addition to this general type of polynomials which can be used to represent any discrete function, a more compact representation of unate Boolean functions is developed. To this end,

- we show that unate Boolean functions can be represented by affine polynomials. As these polynomials grow only linearly in the system size, this is a great advantage over the general case.

- A novel robustness or plausibility measure for these functions is defined. This measure can easily be included into an optimization problem to only search for the most plausible unate update functions.

The performance of all algorithms is tested with a Boolean and a discrete example system, showing the advantage of having the linear programming formulation.

Chapter 4 - Analysis of multistability and multistability robustness

The problems of model validation and model discrimination with respect to a pre-specified multistable behavior is addressed. The first contribution is

- the elaboration of a modeling framework based on ordinary differential equations, in which the model hypothesis and the steady state requirements for the model can be specified adequately.

In this framework, so-called activation and inhibition functions are defined, which are monotonically increasing or decreasing functions, respectively. A hypothesis about the network is formalized by specifying only the monotonicity properties, but not the exact shapes of these functions. Experimentally observed steady states are described as rectangular forward-invariant regions. Their positions in the state space can be specified in qualitative terms by only differing between high and low concentrations. The first methodological contribution is

- an algorithmic solution which allows to decide, if the desired forward-invariant sets can, in principle, be generated by the hypothetical model.

For the second part of this chapter, it is assumed that more quantitative information about the positions of these forward-invariant sets in the state space is available. Based on the same modeling framework,

- the problem of verifying, if a hypothetical model can in principle reproduce the desired forward-invariant regions is formulated as nonlinear feasibility problem. It is proved that this problem is convex for certain system classes.

- A novel robustness measure for the hypothetical model is introduced. Its computation can be realized by adding a convex cost function to the feasibility problem, thus obtaining a convex optimization problem for certain system classes.

The method is applied to several hypothetical models describing a decision step in a model for cell fate decision, identifying sparsity as preferable design principle for these models.

Chapter 5 - Control of Boolean models of gene regulation networks

The problem of controlling Boolean models of gene regulation networks toward an attractor is considered. Thereby, typical limitations on the observability of the system's states and the applicability of inputs are addressed. As algorithmic solutions to this control problem already exist in the discrete event system literature, the contributions of this chapter are as follows.

- We transfer the control problem into a state attraction problem for discrete event systems.

- We show how typical limitations on the applicability of inputs or the observability of states can be model in this framework and included into the state attraction problem.

The method is applied to a Boolean model for cell fate decision and identifies novel control strategies to enforce apoptosis in this system, which could not be found by existing methods.

Chapter 6 - Conclusion

The results and contributions of this thesis are summarized and discussed. Several possible directions of future research are presented.

Appendix

Appendices A and B contain own proofs and examples for Chapters 3 and 4, respectively, to allow a short and coherent presentation in these chapters. Appendix C summarizes two solution approaches for the state attraction problem of discrete event systems known from the literature.

2. Background

While the main part of this thesis deals with theoretical and computational aspects of modeling and analyzing gene regulation network, this chapter gives a comprehensive introduction into the basic biological mechanisms of gene regulation. Two mathematical formalisms to describe gene regulation networks, Boolean or discrete networks, and ordinary differential equations, will be introduced and motivated. Furthermore, a short description of a few measurement techniques to observe the species involved in a gene regulation network is given. Thereby, this chapter does not claim to be complete, but only intends to provide a sufficient background and motivation for the formalisms used in the following chapters.

2.1. Biological fundamentals of gene regulation

All information required for the development and maintenance of an organism is stored in the genome. In higher developed organisms, it can be divided into coding and non-coding DNA. The first type encodes the sequences from which proteins are produced, the second type does not store sequence information for proteins. Among other, this second type however codes for RNAs or other small molecules, which play an important role in regulating biological functions. In each cell of an organism, only a small percentage of the information stored in the genome is processed, depending for example on the developmental stage of the cell, its function or position within the organism, or the intra- and extracellular conditions in general. Reading and processing the genomic information is a highly regulated process, whose proper functioning is essential for a cell to react to varying conditions in an appropriate way.

To describe the most important regulatory mechanisms which are of relevance for this work, let us start with the *central dogma of molecular biology*, which was first stated by Francis and Crick in 1958 (and later re-stated in Crick (1970)). In a simplified form, it states that RNA molecules are produced from DNA, and serve as templates for the production of proteins. The flow of information stored in the DNA sequence is thus from DNA to proteins. The term dogma is however an unfortunate choice and should not be interpreted literally. It rather describes the general transfer of biological sequential information from DNA to DNA (for the replication of cells), from DNA to RNA (transcription), and from RNA to proteins (translation). It does however not preclude special ways of transfer, such as for example reverse transcription from RNA to DNA by retroviruses. We will however only focus on the general transfer.

The transcription of DNA into RNA is carried out by protein complexes called RNA polymerases. In order to initialize the transcription, RNA polymerases usually need to recognize and bind to a certain sequence on the DNA strand, called *promoter*. After binding to that promoter sequence, the DNA double helix is opened and, as the RNA polymerase moves along the DNA strand, the RNA transcript is elongated with ribonucleotides complementary to the DNA sequence. This elongation process is continued until a stop sequence is encountered. In eukaryotes, the RNA is then exported from the nucleus to the cytosol and further modified. There are several types of RNAs. One of these types, the messenger RNA (mRNA), serves as template for the production of proteins. To this end, the mRNA binds, in the cytosol, to ribosomes, which catalyze the translation of the mRNA sequence into a complementary sequence of amino acids. This is done according to a scheme which is highly conserved across most prokaryotic and eukaryotic species, and which assigns each triplet of mRNA bases one amino acid. In post-translational modifications, these polypeptides are then modified or folded to obtain the final and functional protein. These steps are illustrated in Figure 2.1.

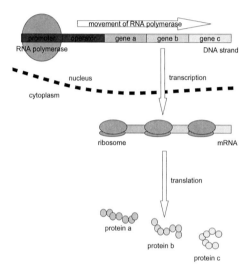

Figure 2.1.: The general transfer of information is from DNA to proteins via the processes of transcription and translation.

Regulation of gene expression takes place at the various steps shortly introduced above, that is, during transcription, processing and transport of the mRNA (both in eukaryotes), and translation. Degradation of mRNAs can play a role, too. Similar to most theoretical works on gene regulation networks, we will however focus on transcriptional control, which is the earliest possible way of influencing gene expression, and which is also considered as the most important type to control gene expression in Klipp *et al.* (2009). In this respect, so-called *transcription factors* play a predominant role. These proteins can bind to specific DNA sequences in the proximity of the promoter sequence, and thus modulate the ability of the promoter to recruit RNA polymerases by either making it easier for RNA polymerases to bind to the promoter, or by blocking the binding sites. The binding of transcription factors to the DNA binding-site motifs is highly specific and can be seen as a feedback from the protein level back to the transcription level, giving rise to a highly complex and interconnected gene regulation network.

2.2. Mathematical modeling of gene regulation networks

A mathematical framework should provide the means to formalize the biological knowledge about the system in a consistent way, in order to obtain a model that can not only reliably reproduce observed behaviors, but that can also be used for predictions. A variety of mathematical modeling frameworks has been proposed to describe the complex relations and interactions between the components of a gene regulation network, comprising graph theory, thermodynamics, logic or Boolean methods, Bayesian methods, and ordinary, partial and stochastic differential equations, to name only the most frequently used ones. More detailed reviews on modeling formalisms for gene regulation networks are for example the works by de Jong (2002), Cho *et al.* (2007), Gardner & Faith (2005), Hecker *et al.* (2009), and Ay & Arnosti (2011). In this work, we will use a Boolean or discrete network description, and a modeling framework based on ordinary differential equations. In the following two sections, these two formalisms are briefly motivated and their advantages and limitations are discussed.

2.2.1. Boolean and discrete models

Boolean models are among the coarsest approaches to describe the dynamic behavior of a gene regulation network. This network description has gained considerable attention since the early works of Kauffman (Kauffman, 1969), but Boolean or similar frameworks have been used even before in the literature (Sugita, 1961, 1963; Walter *et al.*, 1967). In a Boolean model, it is only distinguished if a certain species is present or not. The model has a discrete state space, with state vector $x \in \{0,1\}^n$. Thereby, $x_i(t) = 0$ means that the species represented by x_i is absent at discrete time t, and $x_i(t) = 1$ means that it is present at that point in time. The mutual influences between the species are modeled by Boolean *update functions* $B_i : \{0,1\}^n \rightarrow \{0,1\}$.

These update functions also govern the dynamic evolution of the model. To this end, two qualitatively different approaches have been suggested. If *synchronous* updates are used, a state $x(t+1)$ is computed by evaluating all update rules B_i at the same time, that is, the dynamics follow

$$x_i(t+1) = B_i(x(t)), \quad i = 1, \dots, n. \tag{2.1}$$

This results in a deterministic behavior in the sense that there is a unique successor $x(t+1)$ of a state $x(t)$. As this deterministic sequence of "on" and "off" states is clearly an oversimplification, *asynchronous* updates have been proposed in the literature (Thomas, 1973, 1991; Thomas & D'Ari, 1990; Thomas *et al.*, 1995) in order to introduce more variability into the generation of trajectories. To compute a successor $x(t+1)$ of a state $x(t)$, one has to proceed in two steps.

1. Randomly choose a state x_i.

2. Compute the successor state $x(t+1)$ by setting $x_i(t+1) = B_i(x(t))$ for the state x_i chosen in step 1, and $x_j(t+1) = x_j(t)$ for all other states.

Employing asynchronous updates results in a non-deterministic system behavior, and each state $x(t+1)$ can have at most n different successors. It is also possible to introduce more order into this randomized update process. For example, one can require that all other nodes have to be updated before a specific node can be updated again. Also, a partition into fast and slow nodes has been proposed (Chaves *et al.*, 2005), such that all fast nodes have to be updated before the slow nodes. By a slight abuse of notation, we will use Equation (2.1) independently of the use of synchronous or asynchronous updates.

The most important characteristics of the dynamics of system (2.1) are its attractors. One can easily verify that point attractors are invariant under the chosen update strategy. Attractors consisting of more than one state, as well as the basins of attraction may however change.

An obvious extension of the Boolean framework, yielding a more realistic description of the real process, is to allow more than only two discrete values for each state variable. In a discrete model, the state vector x can take values from the domain

$$\mathcal{D} := \{0, 1, \ldots, d_1\} \times \ldots \times \{0, 1, \ldots, d_n\}. \tag{2.2}$$

The values $d_i \in \mathbb{N}_+$ need not be identical. Each update function B_i is then a mapping from $\mathcal{D}$ to $\{0, 1, \ldots, d_i\}$. Synchronous and asynchronous updates are defined as before, and we will again use Equation (2.1) independently of the update strategy.

Boolean and discrete networks are well-suited to model gene regulation networks in case that there is mainly qualitative knowledge about the biological process. Is not necessary to specify parameter values such as rate constants or affinities. The focus of the Boolean or discrete framework is therefore more on the interaction structure than on details about the reaction kinetics. Yet, despite this simplicity and high level of abstraction, this framework has proved to be complex enough to reproduce many phenomena characteristic for gene regulation networks, such as oscillations or multistability (Kauffman, 1969; Glass & Kauffman, 1973; Thomas, 1973; Thomas & Kaufman, 2001). The Boolean and discrete frameworks have been applied successfully and yielded predictive models in numerous studies (Albert & Othmer, 2003; Sánchez & Thieffry, 2001; Saez-Rodriguez *et al.*, 2007; Chaves *et al.*, 2009; Schlatter *et al.*, 2009; Wittmann *et al.*, 2009b; Kauffman, 1993; Somogyi & Sniegoski, 1996; Bornholdt, 2008; Calzone *et al.*, 2010; Chaouiya & Remy, 2013).

The Boolean and discrete framework will be used in Chapter 3 to develop a method for the identification of Boolean or discrete models of gene regulation networks, and in Chapter 5 to develop control strategies for these systems.

2.2.2. Ordinary differential equations

Ordinary differential equation models of gene regulation networks take the general form

$$\dot{x}_i = f_i(x), \quad i = 1, \ldots, n. \tag{2.3}$$

In this, the state vector $x \in \mathbb{R}_+^n \cup \{0\}$ represents the concentrations of RNAs, proteins, or other small molecules, and the functions $f_i(x) : \mathbb{R}^n \to \mathbb{R}$ describe the molecular

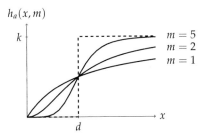

Figure 2.2.: Activating Hill function $h_a(x, m)$ for varying Hill coefficients m. For $m \to \infty$ the function approximates the dashed step function.

effects which govern the dynamic behavior. To determine these right-hand side functions f_i, one can apply essentially the same kinetic laws as for metabolic or signaling processes. The most important such rate laws are the law of mass action, and Michaelis-Menten or Hill type kinetics. The latter follow from the law of mass action under certain quasi steady-state assumptions. If x_i represents an RNA concentration, the right-hand side function f_i often takes the form

$$f_i(x) = -\gamma_i x_i + r_i(x). \tag{2.4}$$

In this, $r_i(x) : \mathbb{R}^n \to \mathbb{R}$ is a regulation function describing how the production rate of x_i is regulated by all other species in the network. Usually, it depends on only few variables. $\gamma_i x_i$ is the degradation term, assuming a constant rate $\gamma_i \in \mathbb{R}_+$. To be more specific about the form of $r_i(x)$, assume that x_j is an activator of x_i. Then, a typical form of $r_i(x)$ is the so-called activating Hill curve (de Jong, 2002; Klipp *et al.*, 2009; Polynikis *et al.*, 2009)

$$r_i(x) = h_a(x_j, m) := \frac{k_i x_j^m}{x_j^m + d_j^m}. \tag{2.5}$$

Similarly, if x_j represses the transcription of x_i, a repressing Hill curve

$$r_i(x) = h_r(x_j, m) := \frac{k_i d_j^m}{x_j^m + d_j^m} \tag{2.6}$$

is assumed. Functions (2.5) and (2.6) are monotonically increasing or, respectively, decreasing functions, in which k_i is the maximum value of $r_i(x)$, m is the so-called Hill coefficient, and d_j can be interpreted as the concentration, at which the regulation function reaches half of its maximum value (see Figure 2.2). If more than one regulator influences the production rate of x_i, it is a common approach to model the joint influence of these regulation factors by sums or products of Hill-type functions. Although Hill-type kinetic functions are only an approximation to describe the kinetics governing the transcriptional dynamics, the sigmoidal shape has been shown to be in good agreement with experimental observations (Yagil & Yagil, 1971).

The obvious advantage of ordinary differential equation models compared to Boolean or discrete models is the continuous state space and the evolution in continuous time.

Yet, more information about the biological processes is needed to specify the forms of the regulation functions and their parameters. If this knowledge is available, deeper insights into the functioning of the system can be gained than with Boolean or discrete models. Therefore, we will base the analysis of multistability and multistability robustness in Chapter 4 on differential equation models as presented in this section.

Let us also point to an interesting connection between the ordinary differential equation and the Boolean or discrete framework. Figure 2.2 illustrates that, for $m \rightarrow \infty$, Hill functions approximate step functions. In this limit, the right-hand side functions $f_i(x)$ become piecewise linear functions. This framework has also been studied extensively in the literature (Glass, 1975; Snoussi, 1989; Casey *et al.*, 2006; de Jong *et al.*, 2004)). It has been shown by Chaves *et al.* (2010), how such a system of piecewise linear functions can be converted into a discrete model with the same qualitative behavior.

2.3. Measurement techniques

Measurements of the species of interest are essential for mathematical modeling. The methods developed in the main part of this thesis always require biological input, either in the form of concrete measurements, or in the form of hypotheses about the model structure. Especially the development of high-throughput methods for the monitoring of mRNA concentrations has enabled the development of mathematical models of gene regulation networks and till today, extensive measurements of mRNA levels are easier and cheaper to obtain than for example measurements of protein concentrations. In this section, we will present a few measurement techniques relevant for the modeling of gene regulation networks.

2.3.1. DNA chips

For the monitoring of mRNA concentrations, DNA chips (also called DNA microarrays) are frequently used. This high-throughput method allows to measure several thousand mRNAs in parallel. For each gene whose expression state is to be measured, fragments of the corresponding DNA sequence are attached to a chip. Then, all mRNAs are extracted from the cell. In a subsequent process called reverse transcription, the mRNAs are transcribed into complementary DNA (cDNA), which is furthermore tagged with fluorescent markers. These cDNA segments are then incubated with the DNA chip, and the cDNAs will bind to the DNA fragments on the chip. Washing off non specific bindings ensures that only specific bindings of cDNAs to their complementary DNA fragments on the chip remain. This will yield a fluorescence signal on the respective spots on the chip. From the positions of these signals, the mRNAs can be identified, while the strength of the signal correlates with the amount of the respective mRNAs in the cell. With this method, only mRNAs with known sequences can be identified. Moreover, only relative signals can be reliably extracted and thus, only fold changes between several measurement can be detected.

2.3.2. ChIP on chip

DNA chips can generate huge amounts of data capturing the expression status of large parts of the genome at certain points in time. However, they provide no further structural information about the complex interactions between genes. Therefore, DNA chips are not suited to predict mutual influences between genes and their products. Thus, if one intends to reconstruct the regulatory interactions of a gene regulation network from DNA chip data alone, the combinatorial complexity of considering each gene product as potential regulator of each other gene quickly becomes intractable. The "ChIP on Chip" (chromatinimmunoprecipitation combined DNA chips) technology allows a significant reduction of this combinatorial complexity as it can predict protein-DNA interactions.

In principal, the ChIP on Chip technology works as follows. Assume we are interested in the DNA binding sites of a specific protein. To this end, DNA and the protein of interest are mixed in an in vitro experiment, such that the protein can bind to its natural binding sites on the DNA. After that, the DNA is fragmented and, with the help of antibodies for the protein of interest, only those DNA fragments are extracted, to which the protein is bound. Protein and DNA are then separated again. In a subsequent DNA chip experiment, similar as described above, these DNA fragments can be identified. The quality of the prediction is limited by the specificity of the antibodies used to extract the protein-DNA complexes and the size of the DNA fragments. It is also an expensive technique.

2.3.3. Fluorescence microscopy

While the above methods require the destruction of the cell, fluorescence microscopy is one way to observe processes in living cells, too. To monitor for example certain proteins, one often uses fluorescent markers, which bind specifically to the protein of interest. Such markers can be antibodies, which are bound to a fluorescent dye. Another approach is the use of the green fluorescent protein (GFP), which has become a standard marker in cell biology. By genetic manipulation, it can be fused with any other protein, such that the cell itself produces a fluorescent variant of the protein. Meanwhile, there even exist several genetic variants of the original GFP which yield signals in different colors. Under a fluorescence microscope, the signals and their strength then allow to draw conclusions about the amount and localization of the respective proteins.

2.3.4. Summary

The measurement techniques briefly introduced above are only a small selection of available techniques to observe mRNA or protein concentrations. There exist numerous other techniques. Quantitative measurements of mRNA concentrations, in contrast to the qualitative measurements obtained by DNA chips, can for example be achieved with real time quantitative PCR (polymerase chain reaction). Protein measurements can also be obtained by blotting techniques or spectrophotometry. The methods described above are however of special importance for the method development in this thesis.

DNA chips are assumed to be the main source of data for the identification and the validation problems addressed in Chapters 3 and 4. The structural information obtained for example from the ChiP on chip technique is used in these chapters, too. It may help to restrict the network to be identified to only those genes and proteins, between which a direct interaction is predicted, or to generate hypotheses about the interaction structure. Fluorescence microscopy is one possibility to obtain measurements from living cells as required for the control problem in Chapter 5. Again, we want to emphasize the large degree of uncertainty inherent to all these measurement techniques. All method development in this thesis will respect this by focusing more on the qualitative than on the quantitative information contained in the data.

3. Identification of Boolean or discrete models of gene regulation networks

In this chapter the problem of identifying discrete models of gene regulation networks is studied. Section 3.1 presents a mathematical formulation of this problem and discusses existing approaches as well as open challenges. In Section 3.2, a polynomial representation of Boolean and discrete functions is developed and its properties are studied. In Section 3.3, this polynomial representation is employed to formulate the problem of finding optimal update functions for the Boolean or discrete model as as a linear program. A Boolean and a discrete application example demonstrate the benefits of the reformulation in Section 3.5. The chapter ends with a summary and discussion in Section 3.6.

Parts of the contents of this chapter have been previously presented in Breindl et al. (2013), covering the representation and estimation of Boolean update functions. Section 3.4 is based on Breindl et al. (2012).

3.1. Introduction and problem statement

The term gene regulation network denotes a collection of interacting DNA segments, mRNAs, proteins and other species. The problem of unraveling the interactions between these species has gained increasing interest over the last one and a half decades (Liang *et al.*, 1998; Akutsu *et al.*, 1999; Kholodenko *et al.*, 2002; Hickman & Hodgman, 2009; Porreca *et al.*, 2012), and has been driven by the advance of high-throughput technologies such as DNA microarrays (see Chapter 2.3), which allow researchers to monitor the activity of thousands of genes in parallel (Gardner & Faith, 2005). Thanks to these technologies, huge datasets can be created. Yet, their exploration to identify the interaction structure still poses severe challenges. Reasons for that are the noisy measurements, the low time resolution, and large number of genes and gene products involved (Polpitiya *et al.*, 2005). Because of these limitations, it is reasonable to focus on the identification of qualitative aspects of the network. With this, we mean, that we do not aim to estimate kinetic details such as affinities or reaction constants, but answer the following elementary questions: Is there an influence from species A on species B? What is the type of this interaction, such as for example, is it activating or inhibiting? How can the coarse data be used to predict the network behavior? As already reasoned in Chapter 2.2, the Boolean, and its generalization to a discrete framework are well suited to answer these questions and will be employed here.

3.1.1. Problem formulation

We consider models with discrete state space that evolve in discrete time. For the sake of convenience, we repeat the model equations from Chapter 2.2, which take the form

$$x_i(t+1) = B_i(x(t)), \quad i = 1, \dots, n. \tag{3.1}$$

Thereby we allow Boolean as well as general discrete models, and synchronous as well as asynchronous updates.

As data for the identification, we consider measurements stemming form high-throughput experiments such as microarrays. It is assumed that all species in the network are measured at several consecutive points in time, and that the data points for each species are mapped to discrete values $x_i \in \{0, 1, \dots, d_i\}$. The number of discretization levels d_i depends on the resolution of the experimental techniques used, and also determines the domain of the variables x_i in model (3.1). We thus assume that the data are arranged in a measurement set

$$\mathcal{O} = \{(x^l, y^l) \mid l = 1, \dots, m\} \tag{3.2}$$

containing m pairs (x^l, y^l), where x and y are the states of (3.1) at two consecutive discrete points in time. That is, each pair satisfies

$$y_i^l = B_i(x^l) \tag{3.3}$$

for all $i = 1, \dots, n$ in the case of synchronous updates, and for one $i \in \{1, \dots, n\}$ in the case of asynchronous updates. All pairs (x^l, y^l) are therefore related by exactly one update step. Also steady state measurements can be described by such a pair for which it holds that $y = x$. The functions B_i are the unknown update functions that have to be identified. Because of the large measurement uncertainties, measurement or discretization errors are considered as well. The measurement set is then

$$\mathcal{O} = \{(\tilde{x}^l, \tilde{y}^l, \mid l = 1, \dots, m\}$$

in which the measured states $\tilde{x}^l$, $\tilde{y}^l$ can differ from the true states x^l, y^l in some components. For notational simplicity, the tilde will be dropped again in the remainder. The problem of finding an update function B_i which can explain as many update pairs as possible can then be formulated as the optimization problem

$$\min_{B: \mathcal{D} \to \{0, \dots, d_i\}} \sum_{l=1}^{m} |B(x^l) - y_i^l|. \tag{3.4}$$

In this, we minimize over all possible discrete functions $B : \mathcal{D} \to \{0, \dots, d_i\}$. The minimizer of (3.4) is then considered as a potential update function B_i for the model. Note however that Problem (3.4) is only correct for the case of synchronous updates. In the case of asynchronous updates, only those pairs (x^l, y^l) should be included into the sum for which it holds that $x_i^l \neq y_i^l$, or if we know that the pair represents a steady state measurement. For notational simplicity, we will however always refer to Problem (3.4).

Restrictions of the search space

The search space of (3.4) grows double exponentially in the number of nodes x_i in the network. In the Boolean case for example, there are 2^{2^n} Boolean functions in n arguments. Because of this explosion on the one hand, and the usually small number of measurements on the other hand, it cannot be expected that a unique minimizer of (3.4) is found. Therefore, some biologically well-motivated assumptions on the update functions are presented next, which lead to a reduction of the search space.

The first assumption concerns the complexity of the update functions. Several studies of well-understood regulation networks in Escherichia coli and Drosophila have revealed that most genes are only influenced by one to three regulators (Leclerc, 2008). It is therefore reasonable to require that each B_i depends on as few arguments as possible.

A second assumption concerns the restriction of the search space of update functions to certain subclasses. Two prominent such subclasses, which have been identified in several studies (Nikolajewa *et al.*, 2007; Raeymaekers, 2002; Wittmann *et al.*, 2010) are the classes unate and (hierarchically) canalizing functions. Their definitions are recalled next.

Definition 3.1 (Unate functions). *An n-dimensional discrete function $B : \mathcal{D} \to \{0, \ldots, d_B\}$, $d_B \in \mathbb{N}_+$, is called unate in x_i, if, for all fixed $(x_1, \ldots, x_{i-1}, x_{i+1}, \ldots, x_n)^T$ with each $x_j \in \{0, \ldots, d_j\}$, and for all $\tilde{x}_i \in \{0, 1, \ldots, d_i - 1\}$, always one of the inequalities*

$$B(\ldots, x_{i-1}, \tilde{x}_i, x_{i+1}, \ldots) \leq B(\ldots, x_{i-1}, \tilde{x}_i + 1, x_{i+1}, \ldots)$$

or

$$B(\ldots, x_{i-1}, \tilde{x}_i, x_{i+1}, \ldots) \geq B(\ldots, x_{i-1}, \tilde{x}_i + 1, x_{i+1}, \ldots)$$

holds. If always the first inequality holds, B is called positive unate in x_i, while it is called negative unate in x_i if always the second inequality holds. B is unate if it is unate in every variable x_i.

While the above definition applies for Boolean as well as for general discrete functions, the (hierarchically) canalizing property has only been defined for Boolean functions.

Definition 3.2 ((Hierarchically) canalizing functions). *A Boolean function $B : \{0, 1\}^n \to \{0, 1\}$ is canalizing in x_i if, for a fixed value $x_i = \hat{x}_i \in \{0, 1\}$, it holds that $B(x) = \hat{B} \in \{0, 1\}$ whenever $x_i = \hat{x}_i$. We call $\hat{x}_i$ the canalizing value of x_i, and $\hat{B}$ the canalized value of x_i. B is called hierarchically or nested canalizing if it is canalizing in all variables x_i together with a hierarchy on these variables as follows: If at $x = \tilde{x}$, several variables x_i take their canalizing value, $B(\tilde{x})$ is defined by the canalized value of the highest variable in the hierarchy.*

Lemma 3.3 (Jarrah *et al.* (2007b)). *If a Boolean function is hierarchically canalizing, then it is unate.*

Several studies have demonstrated the meaningfulness of these definitions for gene regulation networks. An analysis of known regulatory functions in yeast by Harris *et al.* (2002) and Kauffman *et al.* (2003) has shown that the known interactions can indeed be described with (hierarchically) canalizing functions. The use of unate functions was motivated by Grefenstette *et al.* (2006) using a biochemical model of the regulatory mechanisms of gene expression. Also computational or theoretical studies of large random networks have supported these classes. For example, the analysis of random Boolean networks with hierarchically canalizing update functions by Kauffman *et al.* (2004) has revealed that the dynamics are always stable in the sense that perturbations of one node do, in average, not alter the time evolution. The results obtained by Nikolajewa *et al.* (2007) are along the same lines.

In the light of this discussion, Problem (3.4) is modified to obtain the central problem of this chapter

$$\min_{B:\,\mathcal{D}\to d_i} \sum_{l=1}^{m} |B(x^l) - y_i^l|$$

s.t. B has a minimum number of arguments

B is unate or (hierarchically) canalizing.

(3.5)

Even with these two restrictions of the space of update functions, it is still to be expected that uncertainties about the identified interactions remain, such that there will be several minimal models which can explain the observations equally well, and satisfy the unateness or canalizing assumption. It is therefore desirable to be able to formulate additional requirements on the identified network. One reasonable assumption is motivated by the commonly accepted idea, that biological systems are able to perform their respective function in a highly robust way. For gene regulation networks, this can be translated into the requirement that certain steady states are maintained or that a desired trajectory is followed in a stable way despite perturbations. Thus, also the problems of formulating appropriate robustness measures and devising efficient methods for their computation need to be addressed.

3.1.2. Established approaches

Several approaches have been suggested since the late nineties to estimate Boolean or discrete models of gene regulation networks as in Equation (3.1). Some important contributions are briefly discussed next. One of the first contributions, the algorithm REVEAL, has been presented by Liang *et al.* (1998). This algorithm uses information theoretic principles such as Shannon entropy and mutual information analysis to give a "yes" or "no" answer to the question if the temporal evolution of one node can be explained by a given group of nodes. While the algorithm is fast, it does not provide a system description that can be used for simulations. Furthermore, neither measurement errors nor requirements on the update functions as discussed in the previous section can be considered. Another solution has been suggested by Akutsu *et al.* (1999, 2000), which employs an exhaustive enumeration of the search space of Boolean update functions. Because of the exponential growth of this search space, this approach is computationally intractable if the maximal number of arguments of an update functions is not bounded. While these two methods operate on truth tables, the approach followed by Jarrah *et al.* (2007a) and Veliz-Cuba *et al.* (2010) uses a polynomial representation of Boolean and discrete functions. Results from polynomial algebra are applied to construct all possible update functions which can explain a set of observed transition pairs. Also characterizations of hierarchically canalizing and unate cascade functions in an algebraic geometry context are studied. It is however not obvious how measurement errors can be dealt with. Furthermore, the approach requires computationally expensive calculations such as the computation of ideals and their primary decompositions. A further approach which uses a polynomial representation of Boolean functions has been presented by Faisal *et al.* (2008, 2010). The search for Boolean update functions which can explain a set of observed transitions best is formulated as mixed integer quadratic program. Measurement errors are considered, too, and it is studied how the search space of Boolean update functions can be restricted to the class of canalizing functions. The method is however limited to the Boolean setting, and the requirement to solve a mixed integer quadratic program is unnecessarily complex as will be demonstrated in this chapter. As a last approach, the works by Cheng & Qi (2010) and Cheng & Zhao (2011) shall be mentioned. The authors have developed a linear representation of the dynamics of Boolean or general discrete networks. To arrive at this representation, a so-called semi-tensor product is introduced and its properties are studied. A toolbox for Matlab is provided as well. Our experience with this method shows however, that the required intermediate calculations are prohibitively complex, such that moderately sized systems with even less then ten nodes cannot be treated. Also measurement errors cannot be considered. Finally, the development of appropriate robustness measures is a topic that has not been frequently addressed in the literature. As one of the very few approaches, Jarrah *et al.* (2007a) propose several statistical measures to select more plausible models among the various possibilities.

In contrast to these Boolean or discrete approaches, identification methods based on ordinary differential equation (ODE) models are mostly preferred in the control community. Thereby, it is often assumed, that the right-hand side functions are sparse and have certain monotonicity properties which allow the classification of regulators as either activators or inhibitors. For this system class, Porreca *et al.* (2010, 2012) have developed an efficient algorithmic procedure to reject an interaction hypothesis based on the measurement data alone. In a second and refined step, further interaction structures can be excluded with the help of more specific assumptions about the right-hand side functions. Similarly, Cooper *et al.* (2011) have developed a very appealing linear programming formulation to refute hypothetical interaction structures based on the data and a simple, unate-like model for the dynamics of a gene. As reasoned before, we will however use a Boolean or discrete modeling framework in this chapter. A brief comparison with these ODE based approaches will be given in Section 3.3.2. A more general overview over the topic of identifying gene regulation networks can for example be found in the works by Hickman & Hodgman (2009) and Cho *et al.* (2007).

3.1.3. Challenges and goals

The discussion above shows that none of the approaches toward the identification problem offers all of the following desired properties.

- An efficient exploration of the search space of Boolean or discrete functions. Ideally, the formulation should be amenable to standard solvers to profit from their highly optimized design.

- A flexible formulation which is easily extensible to requirements for unateness, the canalizing properties or additional properties such as robustness considerations.

- The result should provide a system description that can be used for predictions.

The goal of this chapter is therefore to formulate a solution to problem (3.5) that exhibits all of these desirable features. To this end, we will

- extend the representation of Boolean functions by polynomials to the general discrete case,

- use this polynomial representation of Boolean and general discrete functions to reformulate problem (3.4) as a standard linear program,

- show that restrictions of the search space can be formulated as constraints to this standard linear program,

- formulate a robustness measure for Boolean models using unate update functions together with an efficient method for its computation.

3.1.4. Notation

Define $N := \{1, 2, \ldots, n\}$ and let ${}^kN$ be the set containing all subsets of N with k different elements. Throughout the chapter, it is assumed that all sets $I \subseteq N$ are ordered, that is, for $I = \{i_1, i_2, \ldots, i_m\}$, it always holds that $i_1 < i_2 < \ldots < i_m$. Next, let x be an n-dimensional vector, and let J be an ordered subset of N. Then, x_J is the $|J|$-dimensional vector with $(x_J)_i = x_{j_i}$, that is, x_J contains only the entries of x which are specified by the set J. The Hamming distance between two Boolean vectors is defined as $H(x, y) := \sum_i |x_i - y_i|$. Finally, the number of discrete vectors in the domain $\mathcal{D}$ of (3.1) is denoted by $d_\Pi := \prod_{i=1}^{n}(d_i + 1)$.

3.2. Polynomial representation of discrete functions

In this section, we address the first challenge of Section 3.1.3, and introduce a polynomial representation of discrete functions B in n arguments, $B : \mathcal{D} \to \{0, \ldots, d_B\}$, with $\mathcal{D} = \{0, \ldots, d_1\} \times \cdots \times \{0, \ldots, d_n\}$ as defined in Section 2.2, and with d_B being a positive natural number. We first define what we mean by a representation of a discrete function by a polynomial.

Definition 3.4 (Polynomial representation of a discrete function). *A polynomial $P : \mathbb{R}^n \to \mathbb{R}$ represents a discrete function $B : \mathcal{D} \to \{0, \ldots, d_B\}$ if it holds that*

$$\forall x \in \mathcal{D} : \ B(x) = P(x). \tag{3.6}$$

The polynomials we propose to use for the representation of discrete functions B take the general form

$$P(x) = \sum_{(r_1, r_2, \ldots, r_n)^T \in \mathcal{D}} a_{r_1, r_2, \ldots, r_n} \, x_1^{r_1} x_2^{r_2} \cdot \ldots \cdot x_n^{r_n}. \tag{3.7}$$

In order to prove that this is a reasonable form, we have to show that all coefficients $a_{r_1,\ldots,r_n}$ can be determined uniquely from the linear system of equations resulting from (3.6). This is first illustrated in an example, and then stated for the general case in Theorem 3.6.

Example 3.5. *Let $n = 2$ and $(d_1, d_2)^T = (2,2)^T$. Then, the polynomial P as in (3.7) is given by*

$$P(x) = a_{0,0} + a_{1,0}x_1 + a_{2,0}x_1^2 + a_{0,1}x_2 + a_{0,2}x_2^2 +$$
$$+ a_{1,1}x_1x_2 + a_{1,2}x_1x_2^2 + a_{2,1}x_1^2x_2 + a_{2,2}x_1^2x_2^2.$$

Given a specific discrete function $B : \{0,1,2\} \times \{0,1,2\} \to \{0,\ldots,d_B\}$, the following system of equations needs to be solved to determine the coefficients of P.

$$
\begin{bmatrix}
1 & 0 & 0 & 0 & 0 & 0 & 0 & 0 & 0 \\
1 & 1 & 1 & 0 & 0 & 0 & 0 & 0 & 0 \\
1 & 2 & 4 & 0 & 0 & 0 & 0 & 0 & 0 \\
1 & 0 & 0 & 1 & 1 & 0 & 0 & 0 & 0 \\
1 & 0 & 0 & 2 & 4 & 0 & 0 & 0 & 0 \\
1 & 1 & 1 & 1 & 1 & 1 & 1 & 1 & 1 \\
1 & 1 & 1 & 2 & 4 & 2 & 4 & 2 & 4 \\
1 & 2 & 4 & 1 & 1 & 2 & 2 & 4 & 4 \\
1 & 2 & 4 & 2 & 4 & 4 & 8 & 8 & 16
\end{bmatrix}
\begin{bmatrix}
a_{0,0} \\
a_{1,0} \\
a_{2,0} \\
a_{0,1} \\
a_{0,2} \\
a_{1,1} \\
a_{1,2} \\
a_{2,1} \\
a_{2,2}
\end{bmatrix}
=
\begin{bmatrix}
B(0,0) \\
B(1,0) \\
B(2,0) \\
B(0,1) \\
B(0,2) \\
B(1,1) \\
B(1,2) \\
B(2,1) \\
B(2,2)
\end{bmatrix}
$$

One can verify that the matrix on the left is invertible. Therefore, all coefficients can be determined uniquely.

In order to show that this linear system of equations has always a unique solution, some preparations are necessary. First, an ordering of the elements x of $\mathcal{D}$ is introduced, which will be used to arrange the coefficients and equations in the linear system of equations in a way which will simplify the proof. To this end, we define a map $\Phi(x) = (I, e)$, such that the ordered set I contains the indices of the nonzero positions of x, and the vector e contains the values of these nonzero positions. More precisely, the set $I = \{i_1, \ldots, i_p\}$, $i_1 < i_2 < \ldots < i_p$, satisfies $x_i \neq 0 \Leftrightarrow i \in I$, and the p-dimensional vector e satisfies $x_{i_l} = k \Leftrightarrow e_l = k$. For example, let $x = (0,3,0,1)^T$. Then, we have $\Phi(x) = (\{2,4\}, (3,1)^T)$. If we are only interested in the set I of $\Phi(x)$, we write $\Phi_I(x)$. Now, let $x, y \in \mathcal{D}$, and let $\Phi(x) = (I, e)$ and $\Phi(y) = (J, f)$. The ordering '$\prec$' of the elements of $\mathcal{D}$ is then defined as follows:

- If $|I| < |J|$, then $x \prec y$.

- If $|I| = |J|$ but $I = \{i_1, \ldots, i_p\} \neq J = \{j_1, \ldots, j_p\}$, then $x \prec y$ if either $i_1 < j_1$ or $i_1 = j_1, i_2 < j_2$, or $\ldots$.

- If $I = J$, then $x \prec y$ if either $e_1 < f_1$ or $e_1 = f_1, e_2 < f_2$, or $\ldots$.

With this, the monomials of P are now arranged such that $a_{r_1,r_2,\ldots,r_n}x_1^{r_1}x_2^{r_2}\cdots\cdot x_n^{r_n}$ appears before $a_{s_1,s_2,\ldots,s_n}x_1^{s_1}x_2^{s_2}\cdots\cdot x_n^{s_n}$ if $(r_1, r_2, \ldots, r_n)^T \prec (s_1, s_2, \ldots, s_n)^T$. Furthermore, the individual equations are arranged such that the equation $P(x) = B(x)$ appears before the equation $P(y) = B(y)$ if $x \prec y$. Note that both, the polynomial P and the system of equations shown in Example 3.5 follow this ordering. In general, the system of equations resulting from Definition 3.4 can be written in matrix form

$$Ea = w, \tag{3.8}$$

in which $a \in \mathbb{R}^{d_{\Pi}}$ is a vector containing the coefficients of P, $w \in \mathbb{R}^{d_{\Pi}}$ is a vector containing all values $B(x)$, $x \in \mathcal{D}$, and $E \in \mathbb{R}^{d_{\Pi} \times d_{\Pi}}$ is a matrix. If the coefficients and equations are arranged as described above, an explicit formula for the matrix E can be given. Let $\chi(k)$ denote the k-th element x in the ordering of $\mathcal{D}$ as described above, then the elements E_{ij} of E are computed as

$$E_{ij} = \begin{cases} \prod_{l \in \Phi_I(y)} x_l^{y_l} & \text{if } \Phi_I(j) \subseteq \Phi_I(i), \text{ and with} \\ & x = \chi(i), \ y = \chi(j) \\ 0 & \text{otherwise.} \end{cases} \tag{3.9}$$

This formula is briefly explained. Consider again Example 3.5, and let us compute Element $E_{9,5}$. To this end, we determine $\chi(9) = x = (2,2)^T$, and $\chi(5) = y = (0,2)^T$. According the formula, it follows that $E_{9,5} = 2^2 = 4$. The first important result concerning the uniqueness of the coefficients of P can now be stated.

Theorem 3.6. *Given an n-dimensional discrete function $B : \mathcal{D} \to d_B$, all coefficients of the polynomial P as in (3.7) are determined uniquely such that P represents B.*

Proof. We have to show that E as in (3.9) is a full rank matrix. From Equation (3.9) (also see Example 3.5) one can see that E is a lower block tridiagonal matrix. To show that E has full rank, it is sufficient to show that all blocks on the main diagonal have full rank. We first examine the blocks comprising all rows i and columns j, such that only the k-th entry of $\chi(i)$ and of $\chi(j)$ is nonzero. According to (3.9), such a block, denoted by B^k, has the form

$$B^k = \begin{bmatrix} 1^1 & 1^2 & \dots & 1^{d_k} \\ 2^1 & 2^2 & \dots & 2^{d_k} \\ \vdots & \vdots & \ddots & \vdots \\ d_k^1 & d_k^2 & \dots & d_k^{d_k} \end{bmatrix}. \tag{3.10}$$

Next, recall that the well-known Vandermonde matrix

$$V(x) = \begin{bmatrix} x_1^0 & x_1^1 & \dots & x_1^{n-1} \\ x_2^0 & x_2^1 & \dots & x_2^{n-1} \\ \vdots & \vdots & \ddots & \vdots \\ x_n^0 & x_n^1 & \dots & x_n^{n-1} \end{bmatrix}$$

has full rank if and only if the values x_i are pairwise distinct. As B^k can be obtained from $V(x)$ by setting $x_i = i$, and multiplying each row with x_i, also B^k has full rank. We next study blocks on the main diagonal comprising rows i and columns j, such only the entries $k \in I = \{i_1, i_2, \dots, i_m\}$, $m > 1$, of the vectors $\chi(i)$ and $\chi(j)$ are nonzero. Splitting I into i_1 and $\bar{I} = \{i_2, \dots, i_m\}$, one can verify that each such block, denoted by B^I, can be expressed as $B^I = B^{i_1} \otimes B^{\bar{I}}$, where $\otimes$ denotes the Kronecker product of matrices (see Example 3.5 and Appendix A.1 for the general form of this matrix). As, for p and q dimensional matrices A and B, it holds that $\det(A \otimes B) = (\det A)^q (\det B)^p$, and as we know that B^I is a full rank matrix if I has only one element, an inductive argument allows to conclude that each block B^I is a full rank matrix. In summary, we have shown that E is a full rank matrix. $\square$

Although P has a large number of monomials, it is still minimal in the following sense. It includes no monomials in which a variable x_i appears with a larger exponent than d_i. This is not necessary as for $x_i \in \{0, 1, \ldots, d_i\}$ and $l > d_i$, x_i^l can always be expressed as a linear combination of $x_i^0, x_i^1, \ldots, x_i^{d_i}$. Also, a simpler polynomial P with less than d_Π monomials would lead to an overdetermined system of equations to determine the coefficients of P.

The Boolean case

In the Boolean case, (3.7) has the form of a multi-affine polynomial

$$P(x) = \sum_{(r_1, r_2, \ldots r_n)^T \in \{0,1\}^n} a_{r_1, r_2, \ldots r_n} x_1^{r_1} x_2^{r_2} \ldots x_n^{r_n}, \tag{3.11}$$

as already used for example by Franke (1994), Wittmann *et al.* (2009a), and Faisal *et al.* (2010). This simpler form allows us to study some properties of P in more detail, and derive analytic expressions for them. The first interesting question concerns the domains from which the coefficients of P can take their values when P represents a Boolean function. For arbitrary Boolean functions, the following result, which is easily obtained, is recalled.

Proposition 3.7 (Ranges of coefficients (Franke (1994))). *The coefficients of P as in (3.11) representing a Boolean function are integers satisfying*

$$a_{0,\ldots,0} \in \{0,1\}$$
$$a_r \in \{-2^{k-1}, \ldots, -1, 0, 1, \ldots, 2^{k-1}\},$$

in which k is the number of nonzero elements in $r \in \{0,1\}^n \backslash (0, \ldots, 0)^T$.

We next analyze the effect of the restrictions to the cases of unate and hierarchically canalizing functions on the ranges of the coefficients. From Equation (3.8) we know that the coefficients are determined by the system of equations

$$a = E^{-1}w, \tag{3.12}$$

which requires the computation of the inverse of E. Therefore, an analytic element-wise expression for E^{-1} is derived next, which is more helpful for the following computations than the recursive definition by Franke (1994). For the Boolean case, it follows from (3.9) that

$$E_{ij} = \begin{cases} 1 & \text{if } \Phi_{I,\chi}(j) \subseteq \Phi_{I,\chi}(i) \\ 0, & \text{otherwise,} \end{cases} \tag{3.13}$$

in which we write $\Phi_{I,\chi}(i) := \Phi_I(\chi(i))$ for better readability.

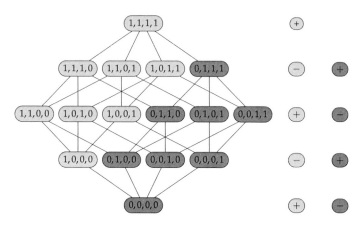

Figure 3.1.: Monotonic ordering of the vertices of the 4-dimensional unit cube depicted as lattice. The $<$-relation is indicated by lines between the nodes. The "$+$" and "$-$" nodes on the right indicate if the values $B(x)$ of the nodes x in the respective level are considered with positive or negative sign for the computation of $a_{1,\dots,1}$. The sublattice for node $(0,1,1,1)^T$ and the respective signs are shown in darker gray.

Lemma 3.8. *In the Boolean case, the inverse of E as in (3.13) is given by*

$$E_{ij}^{-1} = \begin{cases} 1 & \text{if } \Phi_{I,\chi}(j) \subseteq \Phi_{I,\chi}(i) \text{ and } |\Phi_{I,\chi}(i)| - |\Phi_{I,\chi}(j)| \text{ is even,} \\ -1 & \text{if } \Phi_{I,\chi}(j) \subseteq \Phi_{I,\chi}(i) \text{ and } |\Phi_{I,\chi}(i)| - |\Phi_{I,\chi}(j)| \text{ is odd,} \\ 0, & \text{otherwise.} \end{cases} \quad (3.14)$$

The proof of Lemma 3.8 is given in Appendix A.2. We next proceed to determine the maximal absolute value of a coefficient, when w stems from a unate or hierarchically canalizing function.

As basis for the following computations, let us consider the case of monotonic Boolean functions, that is, functions which are positively unate in all arguments. With Lemma 3.8, the computation of a coefficient a_r can be nicely visualized. To this end, the vertices of the n-dimensional unit cube are arranged in a lattice structure such that $(1,\dots,1)^T$ is the supremal element (level n), and $(0,\dots,0)^T$ is the infimal element (level 0). The i-th level contains all nodes $x \in \{0,1\}^n$ which have i nonzero entries, that is, all nodes with $|\Phi_I(x)| = i$. For a node x on level i and a node y on level $i+1$, we define $x < y$ if $\Phi_I(x) \subset \Phi_I(y)$. This is shown for the 4-dimensional case in Figure 3.1.

Consider now the coefficient $a_{1,\dots,1}$. Its computation according to Equation (3.12) can be depicted as follows. Assign each node x in the lattice the value $(-1)^{H((1,\dots,1)^T,x)}B(x)$, where $H(x,y)$ is the Hamming distance between two Boolean vectors x and y. Then, $a_{1,\dots,1}$ is obtained by summing up the values of all nodes in the lattice. The computation of all other coefficients can be illustrated in a similar fashion. Assume, that we want to compute a_r. To this end, consider the sublattice with r as supremal element, denoted by $\mathcal{L}_r$ (see Figure 3.1 for an example). Equation (3.14) states that the same construction as for the computation of $a_{1,\dots,1}$ can now be applied to $\mathcal{L}_r$ to obtain a_r. The maximal absolute value of a coefficient a_r when B is a monotonic Boolean function is given by

$$\hat{a}_r = \max_{\text{monotonic } B} \left| \sum_{y \in \mathcal{L}_r} (-1)^{H(r,y)} B(y) \right|. \tag{3.15}$$

Lemma 3.9. *The maximal absolute value the coefficient $a_{1,\ldots,1}$ of a polynomial P as in (3.11) can assume when P represents a monotonic Boolean function is given by*

$$\hat{a}_{1,\ldots,1} = \binom{n-1}{\hat{\imath}-1}, \tag{3.16}$$

in which

$$\hat{\imath} = \begin{cases} \frac{n}{2} & \text{for } n \text{ even} \\ \frac{n}{2} + \frac{1}{2} & \text{for } n \text{ odd.} \end{cases}$$

The proof of this Lemma is given in Appendix A.3. We can now state the following result.

Proposition 3.10 (Ranges of coefficients for monotonic Boolean functions)**.** *The coefficients of P as in (3.11) representing a monotonic Boolean function are integers satisfying*

$$a_{0,\ldots,0} \in \{0,1\}$$
$$a_r \in \left\{ -\binom{|\Phi_I(r)|-1}{j}, \ldots, -1, 0, 1, \ldots, \binom{|\Phi_I(r)|-1}{j} \right\},$$

where $r \in \{0,1\}^n \backslash (0,\ldots,0)^T$, and $j = \frac{|\Phi_I(r)|}{2} - 1$ if $|\Phi_I(r)|$ is even, and $j = \frac{|\Phi_I(r)|}{2} - \frac{1}{2}$ if it is odd.

Proof. The claim for $a_{0,\ldots,0}$ follows from $B((0,\ldots,0)^T) = a_{0,\ldots,0}$, and Lemma 3.9 proves it for $a_{1,\ldots,1}$. So, it remains to be shown for a_r, $1 \leq |\Phi_I(r)| < n$. From the illustration above, we know however that only the sublattice $\mathcal{L}_r$ and monotonic Boolean functions on this sublattice need to be considered for the computation of the maximal value of $|a_r|$. Application of Lemma 3.9 to polynomial representations of Boolean functions on this sublattice then gives the desired result. $\qquad\square$

Let us next generalize Proposition 3.10 to unate functions. Interestingly, the same result can be stated.

Proposition 3.11 (Ranges of coefficients for unate Boolean functions)**.** *The coefficients of P as in (3.11) representing a unate Boolean function are integers satisfying*

$$a_{0,\ldots,0} \in \{0,1\}$$
$$a_r \in \left\{ -\binom{|\Phi_I(r)|-1}{j}, \ldots, -1, 0, 1, \ldots, \binom{|\Phi_I(r)|-1}{j} \right\},$$

where $r \in \{0,1\}^n \backslash (0,\ldots,0)^T$, and $j = \frac{|\Phi_I(r)|}{2} - 1$ if $|\Phi_I(r)|$ is even, and $j = \frac{|\Phi_I(r)|}{2} - \frac{1}{2}$ if it is odd.

Proof. The result will be proved by relating the unate to the monotonic case. To this end, note that for every unate Boolean function B, there is a (not necessarily unique) transformation of variables $\tilde{x} = T(x)$, defined by

$$\tilde{x}_i = \begin{cases} x_i & i \notin J \\ 1 - x_i & i \in J, \end{cases} \tag{3.17}$$

such that $\tilde{B}(x) := B(T(x))$ is monotonic in x. In this, $J \subseteq N$ contains all indices i such that B is negatively unate in x_i. Next, consider two lattices of nodes, $\mathcal{L}$ and $\tilde{\mathcal{L}}$, whose nodes are related by such a transformation. We first prove an intermediate result: For every node $\tilde{x}$ of level $\tilde{k}$ of $\tilde{\mathcal{L}}$, the node $x = T^{-1}(\tilde{x})$ of $\mathcal{L}$ is either always on an even or always on an odd level of $\mathcal{L}$. We show this by induction over the levels of $\tilde{\mathcal{L}}$. To begin, assume that node $T^{-1}((0,\ldots,0)^\top)$ is on level k of $\mathcal{L}$. Consider a node $\tilde{x}$ of the first level of $\tilde{\mathcal{L}}$, that is, $\tilde{x}$ has only one nonzero entry, $\tilde{x}_j = 1$. From (3.17) we can conclude that $T^{-1}(\tilde{x})$ is either on level $k + 1$ of $\mathcal{L}$ if $j \notin J$, or on level $k - 1$ of $\mathcal{L}$ if $j \in J$. As this holds for every node $\tilde{x}$ of level 1 of $\tilde{\mathcal{L}}$, and as $k + 1$ and $k - 1$ are both either even or odd, the base case is established. Next, assume that the claim is true for all nodes of all levels up to level $\tilde{l}$ of $\tilde{\mathcal{L}}$. Then, for each node $\tilde{x}$ of level $\tilde{l}$ of $\tilde{\mathcal{L}}$, the node $x = T^{-1}(\tilde{x})$ is either always on a level from the set $\{0, 2, 4, \ldots\}$, or always on a level from the set $\{1, 3, 5, \ldots\}$. The same argument as in the base case now shows that for each node $\tilde{x}$ of level $\tilde{l} + 1$ of $\tilde{\mathcal{L}}$, the node $x = T^{-1}(\tilde{x})$ is either always on a level from the set $\{1, 3, 5, \ldots\}$, or always on a level from the set $\{0, 2, 4, \ldots\}$.

Now, let B be represented by a polynomial P as in (3.11). Recall that the value of a coefficient a_r of P is computed as $a_r = \sum_{y \in \mathcal{L}_r} (-1)^{H(r,y)} B(y)$, which means that only nodes $y \in \mathcal{L}_r$, with $\Phi_I(y) \subseteq \Phi_I(r)$, need to be considered for this computation. We then collect all nodes $\tilde{y} \in \tilde{\mathcal{L}}$, which correspond to a node $y \in \mathcal{L}_r$ via the transformation $\tilde{y} = T(y)$, in a set $\tilde{Y}$. It is given by

$$\tilde{Y} = \{\tilde{y} \in \tilde{\mathcal{L}} \mid \tilde{y}_i = 0 \text{ for } i \in \overline{\Phi_I(r)} \backslash J \text{ and } \tilde{y}_i = 1 \text{ for } i \in \overline{\Phi_I(r)} \cap J\},$$

where $\overline{X}$ denotes the complement of a set X in N. Note that $\tilde{Y}$ is a sublattice of $\tilde{\mathcal{L}}$ whose supremal element $\tilde{y}^\top$ satisfies $\tilde{y}_i^\top = 1$ for $i \in \Phi_I(r) \cup J$ and $\tilde{y}_i^\top = 0$ otherwise, and whose infimal element $\tilde{y}^\perp$ satisfies $\tilde{y}_i^\perp = 0$ for $i \in \Phi_I(r) \cup \overline{J}$ and $\tilde{y}_i^\perp = 1$ otherwise. This sublattice is denoted by $\tilde{\mathcal{L}}_{\top,\perp}$. Also note that $\mathcal{L}_r$ and $\tilde{\mathcal{L}}_{\top,\perp}$ have the same number of levels and the same number of nodes in each respective level. It follows

$$a_r = \sum_{y \in \mathcal{L}_r} (-1)^{H(r,y)} B(y) = \sum_{\tilde{y} \in \tilde{\mathcal{L}}_{\top,\perp}} (-1)^{H(r,T^{-1}(\tilde{y}))} \tilde{B}(\tilde{y}).$$

From the intermediate result, it can be concluded that $H(r, T^{-1}(\tilde{y}))$ is either always even or always odd for all nodes $\tilde{y}$ from the same level of $\tilde{\mathcal{L}}$. Therefore, if we are interested only in the absolute value of a_r, we can write

$$|a_r| = \left| \sum_{\tilde{y} \in \tilde{\mathcal{L}}_{\top,\perp}} (-1)^{H(\tilde{y}^\top, \tilde{y})} \tilde{B}(\tilde{y}) \right|$$

as $H(\tilde{y}^\top, \tilde{y})$ is either always even or always odd for all nodes $\tilde{y}$ from the same level of $\tilde{\mathcal{L}}$, too. Next, note that, if we subtract $\tilde{y}^\perp$ from every node in $\tilde{\mathcal{L}}_{\top,\perp}$, the sublattice

$\tilde{\mathcal{L}}_{\tilde{g}^\top - \tilde{g}^\perp}$ is obtained, which has $(0, \dots, 0)^T$ as infimal element but the same structure as $\tilde{\mathcal{L}}_{\top, \perp}$. It holds that

$$\max_{\text{unate } B} |a_r| = \max_{\text{monotone } \tilde{B}} \left| \sum_{\tilde{y} \in \tilde{\mathcal{L}}_{\top, \perp}} (-1)^{H(\tilde{g}^\top, \tilde{y})} \tilde{B}(\tilde{y}) \right|$$

$$= \max_{\text{monotone } \tilde{B}} \left| \sum_{\tilde{y} \in \tilde{\mathcal{L}}_{\tilde{g}^\top - \tilde{g}^\perp}} (-1)^{H(\tilde{g}^\top - \tilde{g}^\perp, \tilde{y})} \tilde{B}(\tilde{y}) \right|.$$

The last equality means however that the maximal value of $|a_r|$ in the case of unate functions is the same as in the case of monotone functions, which concludes the proof. □

As hierarchically canalizing Boolean functions are a subset of the class of unate Boolean functions, the following corollary is immediate.

Corollary 3.12. *Proposition 3.11 also holds for the class of hierarchically canalizing Boolean functions.*

The results obtained in this section and their derivation give some deep insights into the structure of unate Boolean functions and their polynomial representations. Although we will not follow up on this, we have furthermore experienced that, depending on the linear programming solver, some of the optimization problems derived in the following section can be sped up by providing this additional information about the ranges of the coefficients.

3.3. Reformulation of the identification problem as a linear program

We will now make use of the representation of Boolean or general discrete functions by polynomials to achieve the second and the third goal formulated in Section 3.1.3, that is, the reformulation of Problem 3.4 as a linear program, and the formulation of biologically reasonable restrictions as constraints to this optimization problem. As only the problem of finding one update function B_i will be considered, the indices i of Problems (3.4) and (3.5) are dropped in the remainder, when no ambiguities are possible.

3.3.1. Identification of arbitrary discrete functions

Let P be a polynomial as in (3.7), and let a be the vector containing all coefficients of P (arranged according to the ordering introduced in the last section). Furthermore, let $e(x)$ denote the unique row of E, for which it holds that $e(x)a = P(x)$. An immediate reformulation of the combinatorial problem as in Equation (3.4) is then

$$\min_a \sum_{l=1}^m |e(x^l)a - y_i^l| \tag{3.18}$$

$$\text{s.t. } Ea \in \{0, \dots, d_B\}^{d_\Pi}.$$

In this, the integer constraints are necessary to ensure that the polynomial P, with coefficients taken as the minimizer a of (3.18), indeed represents a discrete function B minimizing (3.4). The absolute values in the cost function can be avoided by introducing the new vector of optimization variables $v = (v_1, \ldots, v_m)^T$. One can verify that (3.18) and the mixed integer linear program (MILP)

$$\min_{a,v} \sum_{l=1}^{m} v_l$$
$$\text{s.t.} \ -v_l \leq e(x^l)a - y_i^l \leq v_l, \quad l = 1, \ldots, m \tag{3.19}$$
$$Ea \in \{0, \ldots, d_B\}^{d_\Pi}$$

are equivalent. While there already exist powerful solvers for this class of problems, we intend to find a computationally still more attractive solution. Therefore, the integer constraints are next replaced by a number of inequality constraints to obtain

$$\min_{a,v,t,p} \sum_{l=1}^{m} v_l$$
$$\text{s.t.} \ -v_l \leq e(x^l)a - y_i^l \leq v_l, \quad l = 1, \ldots, m$$
$$t_u \leq (Ea)_u \leq t_u + 1, \quad u = 1, \ldots, d_\Pi \tag{3.20}$$
$$0 \leq p_{uk} \leq 1, \quad u = 1, \ldots, d_\Pi, \ k = 1, \ldots d_B - 1$$
$$t_u = \sum_k p_{uk}, \quad u = 1, \ldots, d_\Pi.$$

In this, the decision variables are the vector of coefficients a, and the vectors $v :=(v_1, \ldots, v_m)^T$, $t := (t_1, \ldots, t_{d_\Pi})^T$, and $p := (p_{11}, \ldots, p_{1(d_B-1)}, \ldots, p_{d_\Pi(d_B-1)})^T$. We furthermore define $X := (v, a, t, p)$. We will show next that certain well characterized optimal solutions of this linear program correspond to optimal solutions of (3.19), and thus, to optimal solutions of (3.4). To this end, the intuition behind the constraints in (3.20) is discussed first. Assume that we can enforce that, for each of the double-inequality constraints, always one of the two possible inequalities is satisfied with equality. Then, it follows that each p_{uk} will either be 0 or 1, and therefore, each t_u will be an integer from the set $\{0, 1, \ldots, d_B - 1\}$. Finally, each entry $(Ea)_u$ of the vector Ea will be an integer from the set $\{0, 1, \ldots, d_B\}$ as desired. It will be a consequence of Lemma 3.13 that it is indeed possible to achieve that always one of the two inequality constraints is satisfied with equality.

Before this lemma is stated, some basic geometric characterizations of the feasible region of (3.20), which is a (convex) polyhedron, are recalled. Consider a general polyhedron $Q = \{z \mid Az \leq b\}$, where $z \in \mathbb{R}^n$, A is a matrix, and b is a vector. Then, a set F is a face of Q if and only if it is nonempty and $F = \{z \in Q \mid A'z = b'\}$, in which $A'z \leq b'$ is obtained from $Az \leq b$ by removing some inequalities (Schrijver (1999), p. 101). The set F is a minimal face of Q if it contains no other faces. According to Theorem 8.4 in Schrijver (1999), it holds that F is a minimal face of Q if and only if $\emptyset \neq F \subseteq Q$ and $F = \{z \mid A'z = b'\}$, with A' and b' as above. Note that, in contrast to the case of non-minimal faces, now only the minimal faces can be solutions to the system of equations $A'z = b'$. All minimal faces have the same dimension $n - \text{rank}(A)$ (Schrijver (1999), p.104).

Lemma 3.13. *The minimal faces of the feasible region of (3.20) are points, that is, vertices. For each vertex X, the following holds: If the entries of the a subvector of X are taken as coefficients of a polynomial P according to (3.7) (and in the right ordering as described in the last section), then P represents a discrete function $B : \mathcal{D} \to d_B$. Also the converse is true. For each discrete function $B : \mathcal{D} \to d_B$, there is at least one vertex X such that the a subvector of X contains the coefficients of a polynomial P according to (3.7) representing B.*

Proof. To show that the minimal faces have dimension zero, we rewrite the constraints of (3.20) in matrix form. Let the matrix $\tilde{E}$ contain all row vectors $e(x^l)$, $l = 1, \ldots, m$, define $y := (y_i^1, \ldots, y_i^m)^T$, and let $\hat{I} = I_{d_\Pi} \otimes (1, \ldots, 1)$ be the Kronecker product of the $d_\Pi \times d_\Pi$ identity matrix with an all-ones row vector of length $d_B - 1$, i.e., $\hat{I}$ results from the identity matrix by replacing each one entry on the main diagonal with the all-ones row vector, and by replacing each zero entry with an all-zeros row vector. Furthermore, $\mathbf{0}$ and $\mathbf{1}$ denote all-zeros and all-ones matrices or vectors of appropriate dimension. With this, the constraints of (3.20) can be written as $AX \leq b$, with A, X and b defined by

$$
\begin{pmatrix}
-I_m & \tilde{E} & \mathbf{0} & \mathbf{0} \\
-I_m & -\tilde{E} & \mathbf{0} & \mathbf{0} \\
\mathbf{0} & E & -I_{d_\Pi} & \mathbf{0} \\
\mathbf{0} & -E & I_{d_\Pi} & \mathbf{0} \\
\mathbf{0} & \mathbf{0} & I_{d_\Pi} & -\hat{I} \\
\mathbf{0} & \mathbf{0} & -I_{d_\Pi} & \hat{I} \\
\mathbf{0} & \mathbf{0} & \mathbf{0} & I_{d_\Pi \cdot (d_B-1)} \\
\mathbf{0} & \mathbf{0} & \mathbf{0} & -I_{d_\Pi \cdot (d_B-1)}
\end{pmatrix}
\begin{pmatrix}
v \\
a \\
t \\
p
\end{pmatrix}
\leq
\begin{pmatrix}
y \\
-y \\
\mathbf{1} \\
\mathbf{0} \\
\mathbf{0} \\
\mathbf{0} \\
\mathbf{1} \\
\mathbf{0}
\end{pmatrix}.
\tag{3.21}
$$

In this, the last equality constraint of (3.20) has been replaced with two inequality constraints. As E has full rank, A has full column rank. It follows that all minimal faces of the polyhedron described by Equation (3.21) are vertices.

Denote the dimension of the vector $X = (v, a, t, p)^T$ by $\hat{n}$. Then A has $2\hat{n}$ rows and $\hat{n}$ columns. In order to determine a vertex, we have to find a vector X with $AX \leq b$ that satisfies a rank $\hat{n}$ subsystem of $AX \leq b$ with equality, i.e., $A'X = b'$. We next show that, whenever X satisfies such a full rank subsystem with equality, it holds that $Ea \in \{0, \ldots, d_B\}^{d_\Pi}$.

As a preparation, let us first collect all inequalities from Equation (3.20) belonging to a given index u in a new system $A^u X^u \leq b^u$. That is, this systems contains all inequalities $t_u \leq (Ea)_u \leq t_u + 1$ as well as $0 \leq p_{u,k} \leq 1$ and $t_u \leq \sum_k p_{u,k} \leq t_u$ for a fixed index u. Furthermore, assume that for the current index u, there are $r^u \geq 0$ tuples in the measurement set, indexed as $(x^{l_{u,j}}, y^{l_{u,j}})$, $j = 1, \ldots, r^u$, which all satisfy $e(x^{l_{u,j}})a = (Ea)_u$. We then also include all inequalities $v_{l_{u,j}} \leq e(x^{l_{u,j}}) \leq v_{l_{u,j}}$ from Equation (3.20) into this new subsystem. As all expressions $e(x^{l_{u,j}})$ refer to the u-th row of E, we replace them by E_u, i.e., the u-th row of E. With $\mathbf{1}_m^n$ denoting the unit row vector of length n with its only 1-entry at position m, the subsystem $A^u X^u \leq b^u$ can be written as

$$
\begin{matrix}
(\alpha_1^u) \\
(\overline{\alpha}_1^u) \\
\vdots \\
(\alpha_{r^u}^u) \\
(\overline{\alpha}_{r^u}^u) \\
(\beta^u) \\
(\overline{\beta}^u) \\
(\gamma^u) \\
(\overline{\gamma}^u) \\
(\delta_1^u) \\
(\overline{\delta}_1^u) \\
\vdots \\
(\delta_{d_B-1}^u) \\
(\overline{\delta}_{d_B-1}^u)
\end{matrix}
\left(
\begin{matrix}
-\mathbf{1}_1^r & E_u & 0 & 0 \\
-\mathbf{1}_1^r & -E_u & 0 & 0 \\
\vdots & \vdots & \vdots & \vdots \\
-\mathbf{1}_r^r & E_u & 0 & 0 \\
-\mathbf{1}_r^r & -E_u & 0 & 0 \\
0 & E_u & -1 & 0 \\
0 & -E_u & 1 & 0 \\
0 & 0 & 1 & -(1,\dots,1) \\
0 & 0 & -1 & (1,\dots,1) \\
0 & 0 & 0 & \mathbf{1}_1^{d_B-1} \\
0 & 0 & 0 & -\mathbf{1}_1^{d_B-1} \\
\vdots & \vdots & \vdots & \vdots \\
0 & 0 & 0 & \mathbf{1}_{d_B-1}^{d_B-1} \\
0 & 0 & 0 & -\mathbf{1}_{d_B-1}^{d_B-1}
\end{matrix}
\right)
\left(
\begin{matrix}
\begin{pmatrix} v_{l_{u,1}} \\ \vdots \\ v_{l_{u,r^u}} \end{pmatrix} \\
a \\
t_u \\
\begin{pmatrix} p_{u,1} \\ \vdots \\ p_{u,d_B-1} \end{pmatrix}
\end{matrix}
\right)
\leq
\begin{pmatrix}
y_i^{l_{u,1}} \\
-y_i^{l_{u,1}} \\
\vdots \\
y_i^{l_{u,r^u}} \\
-y_i^{l_{u,r^u}} \\
1 \\
0 \\
0 \\
0 \\
0 \\
1 \\
\vdots \\
1 \\
0
\end{pmatrix} .
$$

The matrix A^u only contains the rows and columns from the larger matrix A as in Equation (3.21) which are relevant for the current index u. Each row ξ of A corresponds to a unique row ξ^u in a unique matrix A^u: ξ^u can be obtained from ξ by removing all columns which belong to some variable contained in X but not in X^u. Thereby only zero entries are removed. Similarly, a row ξ of A can be recovered from a row ξ^u of A^u by filling in zeros for all variables contained in X but not in X^u. We denote this relation by the mapping $\xi = A(\xi^u)$, i.e., a row ξ^u in A^u is mapped to a unique row ξ in A. Using this notation, observe that any two rows $A(\xi^u)$ and $A(\zeta^u)$ of A are linearly dependent (independent) if ξ^u and ζ^u of A^u are linearly dependent (independent). Moreover, any two rows $A(\xi^u)$ and $A(\xi^{\hat{u}})$, with $u \neq \hat{u}$, are linearly independent. Let us next study two properties of the matrix A^u.

i) For each choice of $j, k \in \{1, \dots, r^u\}$, the rows $(\alpha_j^u), (\overline{\alpha}_j^u), (\alpha_k^u)$ and $(\overline{\alpha}_k^u)$ are linearly dependent, which follows from $[(\alpha_j^u) + (\overline{\alpha}_j^u)] - [(\alpha_k^u) + (\overline{\alpha}_k^u)] = 0$.

ii) Consider the rows (α_j^u) and $(\overline{\alpha}_j^u)$, $j \in \{1, \dots, r^u\}$, and choose, from all rows of types β to δ, always only one of the two alternative rows, i.e., either the one with bar or the one without bar. Then, the resulting rows are always linearly dependent, as can for example be seen from the fact that $(\alpha_j^u) - (\overline{\alpha}_j^u) - 2(\beta^u) - 2(\gamma^u) - 2(\delta_1^u) - \dots - 2(\delta_{d_B-1}^u) = 0$. Also note that the remaining rows are linearly independent again as soon as one of these rows is removed.

With this, we can now study the properties of solutions X, which satisfy a full rank subsystem of $AX \leq b$ with equality, i.e., $A'X = b'$. To this end, we again consider all rows belonging to a given index u at the same time, i.e., we consider a matrix A^u. Denote the number of rows in A^u by $\hat{n}^u$. We then enumerate all possibilities to choose, from that matrix A^u, exactly $\frac{\hat{n}^u}{2}$ linearly independent rows ξ^u, whose correspondent rows $\xi = A(\xi^u)$ in A are then added to A'. After that, we show that this procedure indeed enumerates all possibilities to build full rank systems $A'X = b'$.

To start, we distinguish two cases for r^u for the current index u.

Case A) $r^u = 0$, i.e., there is no tuple (x^l, y^l) in the measurement set for which $e(x^l) = E_u$, and A^u therefore has no rows of type α. For such u, always include one of the two rows $A(\xi^u)$ or $A(\overline{\xi}^u)$ for each $\xi^u \in \{\beta^u, \gamma^u, \delta_1^u, \ldots, \delta_{d_B-1}^u\}$ into A'. Also observe that there is no other way of choosing $\frac{\hat{n}^u}{2}$ linearly independent rows from A^u.

From the chosen rows it always follows that each $p_{u,k}$ is either 0 or 1, and that t_u is therefore an integer from 0 to $d_B - 1$. Finally, from $A(\beta_u)$ or $A(\overline{\beta}_u)$ it follows that $E_u a = (Ea)_u$ is an integer from 0 to d_B as required.

Case B) $r^u \geq 1$, i.e., the matrix A^u now has rows of type α. From property i) we know that for at most one $j \in \{1, \ldots, r^u\}$, both rows $A(\alpha_j^u)$ and $A(\overline{\alpha}_j^u)$ can be included into a potential full rank system, as we would have linearly dependent rows in A' otherwise. From ii) we also know that, if for some j, both rows $A(\alpha_j^u)$ and $A(\overline{\alpha}_j^u)$ are in A', then we cannot include at the same time always one of the two rows $A(\xi^u)$ or $A(\overline{\xi}^u)$, with $\xi^u \in \{\beta^u, \gamma^u, \delta_1^u, \ldots, \delta_{d_B-1}^u\}$. These observations lead to the following two subcases, which describe all possibilities to choose $\frac{\hat{n}^u}{2}$ linearly independent rows in A^u.

Case Ba) For each $j \in \{1, \ldots, r^u\}$, always only one of the two rows $A(\alpha_j^u)$ or $A(\overline{\alpha}_j^u)$ is included in A'. We furthermore include always one of the two rows $A(\xi^u)$ or $A(\overline{\xi}^u)$ for each $\xi^u \in \{\beta^u, \gamma^u, \delta_1^u, \ldots, \delta_{d_B-1}^u\}$. With the same reasoning as in case A), it follows that $E_u a = (Ea)_u$ will be an integer from 0 to d_B as required.

Case Bb) For exactly one $j \in \{1, \ldots, r^u\}$, both rows $A(\alpha_j^u)$ and $A(\overline{\alpha}_j^u)$ are included in A'. From these two rows alone we necessarily have $v_{l_{u,j}} = 0$ and $E_u a = (Ea)_u = y^{l_{u,j}}$, which is an integer from the set $\{0, \ldots, d_B\}$ as required.

To obtain $\frac{\hat{n}^u}{2}$ linearly independent rows in A' from the current system A^u, we furthermore include, for all remaining indices $k \in \{1, \ldots, r^u\}$, always one of the two rows $A(\alpha_k^u)$ or $A(\overline{\alpha}_k^u)$ into A'. Finally, we choose always one of the two rows $A(\xi^u)$ or $A(\overline{\xi}^u)$, with $\xi^u \in \{\beta^u, \gamma^u, \delta_1^u, \ldots, \delta_{d_B-1}^u\}$ and include all but one of them into A'.

From this procedure we obtain full rank systems $A'X = b'$ with $\hat{n}$ equations, the solutions of which always satisfy $Ea \in \{0, \ldots, d_B\}^{d_\Pi}$. We finally need to show that the proposed ways to construct such full rank subsystems are indeed the only possibilities.

That this is indeed the case can be seen from the following considerations. First, there are no other ways of choosing $\frac{\hat{n}^u}{2}$ linearly independent rows for a given index u. We also cannot include more rows from that index u into A' than those described above, as otherwise we would always obtain linearly dependent rows in A'. On the other hand, if we wanted to include fewer rows for a given index u than those described above, we would have to include more rows belonging to some other index $\hat{u}$. Otherwise, we would not obtain a square matrix A'. However, including more rows belonging to that index $\hat{u}$ than those described above would again yield linearly dependent rows in A'.

In summary, we have shown that the a subvector of each vertex X contains the coefficients of a polynomial P as in (3.7) such that P represents a discrete function $B : \mathcal{D} \to d_B$. From the above explanations one can also see that for each vector

$\hat{w} \in \{0, \dots, d_B\}^{d_\Pi}$, there is at least one vertex for which a is determined by $Ea = \hat{w}$. Therefore, also the converse direction is true. $\qquad\square$

Lemma 3.13 shows that Problem (3.20) has a nice structure and is relatively easy to solve. From a computational perspective, it allows the use of simplex solvers, which can be expected to be perform with linear complexity in the problem dimensions (Schrijver, 1999).

Corollary 3.14. *Problem (3.20) can be solved with a linear programming solver based on the simplex method. This yields an optimal vertex whose a subvector contains the coefficients of a polynomial P according to (3.7) such that P represents a discrete function B. This discrete function B is a minimizer of (3.4).*

3.3.2. Unate and hierarchically canalizing functions

The unateness or canalizing constraints of Problem (3.5) are no longer ignored now. The goal of this section is to show that these restrictions can be described by additional constraints that do not destroy the beneficial property that the optimal solutions of interest are among the vertices spanning the feasible region.

Let us first study the unateness property and assume that, for each argument x_i, it is known if B is positive or negative unate in x_i. It follows directly from Definition 3.1 that the coefficients a of a polynomial P as in (3.7) representing B have to satisfy a system of inequalities

$$(E_a - E_b)a \le 0, \tag{3.22}$$

in which E_a and E_b are matrices containing rows of E. Each inequality of (3.22) corresponds to one inequality from Definition 3.1. Therefore, $(E_a - E_b)$ will have $\prod_{i=1}^{n}((d_1 + 1) \dots (d_{i-1} + 1) \cdot (d_{i+1} + 1) \cdot \dots \cdot (d_n + 1))d_i$, or, in the Boolean case, $n\,2^{n-1}$ rows.

For the case of canalizing Boolean functions, assume that B is canalizing in x_i with canalizing value $\hat{x}_i$. Then, it has to hold that

$$B(x) - B(y) = 0 \text{ whenever } x_i = y_i = \hat{x}_i.$$

For one canalizing variable, there are $2^{n-1} - 1$ such equality constraints. This can also be expressed by a linear system of inequality constraints for the vector of coefficients a of the form (3.22) (and by replacing each equality by two inequality constraints). If there is a hierarchy of several canalizing variables, more constraints of the same type are necessary: for the k-th variable in the hierarchy, $2^{n-k} - 1$ equality constraints are required.

As both requirements can be formulated in a similar form, we next study the effects of adding constraints (3.22), stemming either from a unateness or canalizing assumption, to the linear program (3.20). We can state the following variation of Lemma 3.13.

Lemma 3.15. *The feasible region of (3.20) with additional constraints (3.22) is a polyhedron generated by a subset of the vertices generating the polyhedron described by the constraints of (3.20) alone. For each vertex X from this subset, the following holds: If the entries of the a subvector of X are used as coefficients of a polynomial P as in (3.7) (and in the right ordering),* then P represents a discrete function $B : \mathcal{D} \to d_B$, which is unate or canalizing as expressed by the additional constraints. Also the converse is true: For each discrete function $B : \mathcal{D} \to d_B$ with the desired unateness or canalizing property, there is at least one vertex X in this subset of vertices such that the a subvector of X contains the coefficients of a polynomial P as in (3.7) representing B.

Proof. We first show that no new vertices are introduced by the additional constraints. Therefore, first note that the matrix $(E_a - E_b)$ cannot have full rank. If (3.22) expresses unateness constraints, this can be seen with the help of an example: Let $a \neq \mathbf{0}$ contain the coefficients of a polynomial P as in (3.7) with $P(x) = const \in \{0, \ldots, d_B\}$ for all $x \in \mathcal{D}$. Then, this vector a satisfies $(E_a - E_b)a = 0$. As $a \neq \mathbf{0}$, $(E_a - E_b)$ cannot have full rank. In the case that (3.22) expresses canalizing constraints, a similar argument leads to the same conclusion.

Let us next examine an arbitrary full rank subsystem of (3.20) with additional constraints (3.22). Resolving this system for the a components leads to

$$
\begin{pmatrix}
(E_a - E_b)_{l_1} \\
\vdots \\
(E_a - E_b)_{l_r} \\
E_{j_1} \\
\vdots \\
E_{j_s}
\end{pmatrix}
a =
\begin{pmatrix}
0 \\
\vdots \\
0 \\
g_1 \\
\vdots \\
g_{s'}
\end{pmatrix},
$$

where each row $(E_a - E_b)_l$ is a row from $(E_a - E_b)$, and each row E_j is a row from E. Also, $g_1, \ldots, g_s \in \{0, \ldots, d_B\}$, and $s \geq 1$ because $(E_a - E_b)$ does not have full rank. Thus, for all rows E_j, $j \in \{j_1, \ldots, j_s\}$, we obviously have $E_j a \in \{0, \ldots, d_B\}$. But $E_p a \in \{0, \ldots, d_B\}$ is also true for all rows $p \notin \{j_1, \ldots, j_s\}$. This follows from the structure of $(E_a - E_b)$ in the unate as well as in the canalizing case: each such row E_p can be obtained as $E_p = E_j + \sum_{q \in Q}(E_a - E_b)_q$ for some index $j \in \{j_1, \ldots, j_s\}$ and index set $Q \subseteq \{l_1, \ldots, l_r\}$. To conclude, all points that satisfy a full rank system of equality constraints of problem (3.20) with additional constraints (3.22), also satisfy a full rank system of equality constraints of problem (3.20) alone. On the other hand, not all vertices spanning the feasible region of problem (3.20) will satisfy the additional constraints (3.22). Therefore, the feasible region of the extended problem is generated by the subset of vertices of the original feasible region which additionally satisfy constraints (3.22). A similar argumentation to the one in the proof of Lemma 3.13 then shows the claims in Lemma 3.15. $\qquad\square$

Corollary 3.16. *Problem (3.20) with additional constraints (3.22) can be solved using a linear programming solver based on the simplex method, while the conclusions of Corollary 3.14 remain true.*

In the Boolean case, the linear problem (3.20) takes the much simpler form

$$
\min_a \sum_{l=1}^{m} \left((-1)^{y_i^l} (e(x^l)a - y_i^l) \right)
$$

$$
\text{s.t. } 0 \leq (Ea)_u \leq 1 \qquad\qquad u = 1, \ldots, 2^n.
$$

(3.23)

The additional vector of decision variables v as used in Problems (3.19) and (3.20) is not necessary here. Thus, for the Boolean case, the linear reformulation even has the same number of optimization variables as the corresponding mixed integer linear program (3.19).

A brief performance study and comparison with ODE based approaches

The linear program (3.20) grows exponentially in the number of arguments of an update function B_i, which restricts the approach according to our experience to approximately 14 variables in the Boolean case. Thereby, the limiting factor is not the computing time, but the memory requirements to store all constraints. The problems however grow only linearly in the number of measurements, and the overall optimization problem is a linear program. Moreover, for approximately up to 5 or 6 regulators, each having two to four discretization levels, this exponential growth imposes no severe limitations. The formulated optimization problems have therefore reasonable sizes under these biologically well motivated restrictions.

The optimization problems formulated for example by Cooper *et al.* (2011) and Porreca *et al.* (2010, 2012) in an ODE setting have the advantage that there is no such explosion concerning the number of regulators per gene. However, the problems grow quadratically with the number of measurements, and the overall problem in Cooper *et al.* (2011) is a quadratic optimization problem. Therefore, the advantage of our approach becomes mainly effective when there are sufficiently many data, which are hard to quantify, and few discretization levels. Then, the exploration of the data according to Cooper *et al.* (2011) or Porreca *et al.* (2010, 2012) becomes more expensive than our approach, where a whole update function is considered. While, in the case of few and well quantifiable data, the approaches by Cooper *et al.* (2011) and Porreca *et al.* (2012) might be preferred, the assumption of many and discretized data is however well motivated, too.

Next, we briefly compare the running times of the mixed integer linear and the linear formulation using the CPLEX solvers from the IBM ILOG suite to demonstrate the advantage of the linear reformulation in Equation (3.20). We do not include algorithms into this study that directly solve problem (3.5), for example by complete enumeration, as we are not aware of efficient formulations operating directly on Boolean or discrete functions. It is however easy to see that enumeration algorithms are not a feasible approach to this problem, as all 2^{2^n} Boolean functions (in the Boolean case) on n dimension have to be enumerated an tested in the worst case, that is, the complexity grows double exponentially in the problem dimension.

Figure 3.2 shows the CPU times as measured by Matlab to solve Problems (3.19) and (3.20) for various randomly generated measurement sets $\mathcal{O}$ and various dimensions. The detailed setup is explained in the caption of this figure. One can see that the linear formulation always performs better than the mixed integer formulation, showing the advantage of the reformulated problem. Especially in the discrete case, this is not an obvious result, as the linear formulation (3.20) requires more optimization variables than the mixed integer formulation (3.19).

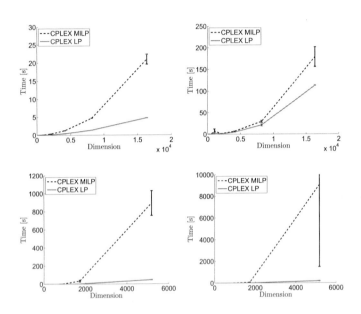

Figure 3.2.: Runtimes of the mixed integer and the linear formulation. On each horizontal axis, the number of elements in $\mathcal{D}$, that is, the number of coefficients of P to be estimated, is plotted. Upper left figure: Optimization over Boolean functions. Upper right figure: Optimization over unate Boolean functions. Lower left figure: Optimization over discrete functions. Lower right figure: Optimization over unate discrete functions. For each data point several optimization runs were performed using randomly chosen data sets of varying sizes.

3.3.3. A solution to the identification problem

Algorithm 3.1 shows how the reformulation of Problem (3.5) can be used to approach the identification problem. Thereby, the linear optimization problem is iteratively solved for an increasing number of regulators up to a maximal number k_{max}, or until a combination of k regulators has been found that can explain the observed transitions in the measurement set $\mathcal{O}$ sufficiently well. For a fixed set of regulators, the optimization is repeated for each assignment of activators and inhibitors. An equivalent algorithm can of course also be given for the case of (hierarchically) canalizing functions.

Algorithm 3.1: Find optimal unate update functions $B_i(x)$ for x_i.

1 **foreach** $k = 1, 2, \ldots k_{max}$ **do**
2   **foreach** $J \in^k N$ **do**
3   **foreach** *assignment of the nodes in x_J as activators or inhibitors* **do**
4   solve Problem (3.20) with Constraints (3.22)
5   **if** *cost function is 0 or below some threshold* **then**
6   save J, the assignment of the nodes in x_J, and the optimal B_i
7   **if** *cost function 0 or below some threshold has been achieved by some J* **then**
8   stop

As the linear program most likely needs to be solved a large number of times, its lower computational complexity as demonstrated in the last section is of great advantage. In Algorithm 3.1, not the complete identification problem is reformulated as a linear program. The outer loops involving the enumeration of all combinations of regulators and of all assignments of activators or inhibitors remain a combinatorial problem. To directly include the search for a minimal number of regulators into the linear program, one can apply regularization techniques. The effectiveness of l_0 regularization and its common l_1 relaxation is next studied in a Boolean setting.

Estimation of minimal Boolean functions

As mentioned above, the update function B_i for a node x_i will most likely not depend on the complete state vector, but only on a small subset of the entries of x. Only very few genes in a large network will depend on a larger number of regulators.

Instead of solving the optimization repeatedly for all possible sets of regulators as in Algorithm 3.1, one can add a regularization term to the cost function. An l_0 regularization term which aims at minimizing the number of arguments that will enter the estimated Boolean functions, can be formulated as a function of the vector of coefficients a,

$$T_0(a) = \left[\left([a^1]_0, [a^2]_0, \ldots, [a^n]_0 \right)^T \right]_0.$$

In this, $[\cdot]_0$ is the l_0-norm of a vector, that is, its number of nonzero entries, and a^j is a vector containing all coefficients a_x of a, for which x_i is nonzero. To see, why we use the nested application of l_0-norms, consider the Boolean function $B(x_1, x_2) = x_1$ and x_2, which is represented by $P(x) = a_{1,1}x_1x_2$. For this vector of coefficients, we have

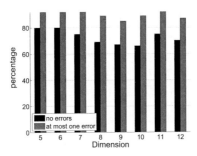

Figure 3.3.: Performance of the l_1 regularization. Black bars: Percentage of experiments in which the estimated minimal number of variables was the true minimal number of arguments. Gray bars: Percentage of experiments, in which the estimated minimal number of arguments differed by not more than one from the real minimal number.

$T_0(a) = [([a_{0,0}]_0, [(a_{1,0}, a_{1,1})^T]_0, [(a_{0,1}, a_{1,1})^T]_0)^T]_0 = 2$, while $[a]_0 = 1$. Thus, for a Boolean function $B(x)$ and its polynomial representation $P(x)$ according to (3.11), $T(a)$ will be equivalent to the number of non-redundant arguments of B. Adding this regularization term, Problem (3.23) becomes

$$\min_a \sum_{l=1}^{m} \left((-1)^{y_i^l} (e(x^l)a - y_i^l) \right) + f T_0(a) \tag{3.24}$$
$$\text{s.t. } 0 \leq (Ea)_u \leq 1 \qquad u = 1, \dots, 2^n,$$

in which $f \in \mathbb{R}_+$ is a weighting factor. However, as T_0 is nonlinear and not even continuous in a, all advantages of the polynomial representation such as its tractability by linear programming methods are lost using the l_0 regularization.

Fortunately, the l_0 minimization problem is a well studied problem. A common relaxation is the replacement of the l_0- by the l_1-norm. Especially in the field of compressive sensing (Donoho, 2006), l_0 minimization and its relaxation by the l_1-norm has gained much attention. It has even been shown that for the reconstruction of sparse signals from compressive measurements, the l_1 relaxation is exact under certain mild requirements (Candès *et al.*, 2006). In Candès *et al.* (2008), one can find a comprehensive summary of other fields, where the l_1 relaxation is successfully applied. In our case, this relaxation leads to

$$T_1(a) = [[a^1]_1, [a^2]_1, \dots, [a^n]_1]_1^T = \sum_{x \in \{0,1\}^n} |a_x| |\Phi_I(x)|,$$

in which $[\cdot]_1$ is the l_1 norm of a vector, and $|\Phi_i(x)|$ is the number of nonzero entries of the Boolean vector x. Problem (3.24), with $T_1(a)$ instead of $T_0(a)$, can then be rewritten as a standard linear program. The performance of the l_1 regularization was then tested in a numerical study. Thereby, 100 experiments were performed for each of several dimensions. For each experiment, a random Boolean function with at most three non-redundant arguments was created and only at most 20 randomly selected pairs $(x, (B(x))$ were used for the estimation. Figure 3.3 shows that, although the true minimal number of arguments $B(x)$ is not always recovered correctly, the regularization performs well and reconstructs a minimal representation in most cases.

3.4. Reduced order representation and robust estimation of unate Boolean functions

The representation of discrete functions by polynomials, and the linear formulation of the identification problem developed in the last two sections are most general in the sense that they can be applied to represent or estimate any discrete function. The disadvantage is however the large number of coefficients of the representing polynomial, and thus the large memory requirement for the resulting optimization problems. As unate Boolean functions are of special interest for the modeling of gene regulation networks, we now develop a more compact representation of these functions, which will also lead to optimization problems with reduced memory requirements.

As already outlined in the introduction, the number of tuples in $\mathcal{O}$ may be too small to derive a unique minimal model. As a second extension, and building on this reduced representation, we will therefore develop a plausibility function to reduce this uncertainty by choosing only the biologically most plausible combinations of regulators for a node x_i.

3.4.1. Reduced order representations for unate Boolean functions

The first step toward a reduced order representation is to replace Definition 3.4 by the weaker notion of a sign-representation (Saks (1993)).

Definition 3.17 (Sign-representation). *Let B be a Boolean function $B : \{0,1\}^n \to \{0,1\}$, and let $P : \mathbb{R}^n \to \mathbb{R}$ be a polynomial. P is a sign-representation of B if $B(x) = \tilde{\mathrm{sgn}}P(x)$ holds for all $x \in \{0,1\}^n$, with $\tilde{\mathrm{sgn}}$ defined by*

$$
\begin{aligned}
\tilde{\mathrm{sgn}}(P(x)) &= 0 &\Leftrightarrow\quad P(x) < 0\\
\tilde{\mathrm{sgn}}(P(x)) &= 1 &\Leftrightarrow\quad P(x) > 0\\
\tilde{\mathrm{sgn}}(P(x)) &= 0.5 &\Leftrightarrow\quad P(x) = 0.
\end{aligned}
$$

This sign-representation can also be seen as threshold function with threshold zero (Muroga, 1971). The following result establishes that the classes of unate Boolean functions and threshold functions coincide for $n \leq 3$.

Theorem 3.18. *Let $B : \{0,1\}^n \to \{0,1\}$, $n \leq 3$, be a Boolean function. B is unate if and only if it can be sign-represented by an affine polynomial $P : \mathbb{R}^n \to \mathbb{R}$, such that $P(x) = a_0 + a^T x$.*

One can verify this result by complete enumeration of all Boolean functions in up to three arguments, but an analytic proof is given by Breindl et al. (2012). Also note, that the restriction $n \leq 3$ in Theorem 3.18 is important. Considering sufficiency, one can easily verify that each Boolean function given by $B(x) = \tilde{\mathrm{sgn}}(a_0 + a^T x)$, $a_0 \in \mathbb{R}$, $a \in \mathbb{R}^n$, with n arbitrary, is unate. The converse direction is however not true in general. One can find counterexamples such that $B : \{0,1\}^n \to \{0,1\}$, $n > 3$, is unate, but no affine sign-representation exists.

A reduced optimization problem to identify unate Boolean functions

Despite the lack of sufficiency of Theorem 3.17 for $n > 3$, we propose to use the sign-representation by affine functions to identify unate Boolean functions, that can explain all transition pairs in $\mathcal{O}$. The optimization problem to find a unate update function for node x_i with arguments x_J, $j \subseteq N$, is then given by

$$
\begin{aligned}
\text{find} \quad & a_0 \in \mathbb{R}, \ a \in \mathbb{R}^{|J|} \\
\text{s.t.} \quad & \forall l \in \{1, \dots m \mid y_i^l = 1\}: \quad a_0 + a x_J^l > 0 \\
& \forall l \in \{1, \dots m \mid y_i^l = 0\}: \quad a_0 + a x_J^l < 0.
\end{aligned}
\tag{3.25}
$$

The following result can be stated immediately.

Lemma 3.19. *For $|J| \leq 3$, Problem (3.25) is feasible if and only if there exists a unate Boolean function $B(x_J)$ that can explain all transition pairs in $\mathcal{O}$. For $n > 3$, such an update function exists if the problem is feasible. If it is infeasible, the existence of such an update function cannot be excluded.*

The disadvantage of this formulation is, that no measurement errors can be considered, that is, only Boolean functions that can reproduce all transition pairs, are found. Furthermore, we give up on generality for more then three dimensions. The advantage is however a significant complexity reduction. Problem 3.25 grows only linearly in $|J|$ compared to the exponential growth of for example Problem (3.23). The example later in this chapter will show that this advantage outweighs the loss of generality.

3.4.2. Identification of robust unate Boolean models

In the introduction of this chapter, the formulation of additional requirements on the identified model considering robustness measures has been stated as a further important goal. For the special case of unate Boolean functions, we will now develop such measures, and show how they can be computed efficiently.

Our development of a robustness measure follows the widely accepted idea that gene regulation networks are able to perform their respective function in a highly robust way and builds on the assumption, that the essential dynamics of the networks, for example the steady states or relevant transients, have been observed and recorded in the measurement set. We therefore require that an update function B_i should produce the same truth values for all observed inputs x^l, $l = 1, \dots, m$, even if the underlying biological mechanisms are perturbed.

In order to find a way to describe this ability with a robustness measure $\mathcal{R}$, consider the following geometric illustration of Theorem 3.18. Let B be a three-dimensional Boolean function, and let each value $x \in \{0,1\}^3$ be associated with a vertex of the three-dimensional unit cube. These vertices are colored according to the value of $B(x)$ in the following way. If $B(x) = 0$, the respective vertex is colored black, if $B(x) = 1$, it is colored gray (see Figure 3.4). Then, Theorem 3.18 states that the black and the gray vertices can be strictly separated by a plane only if B is unate. A very intuitive idea is now to use the distance of the separating $(n-1)$-dimensional hyperplane to the nearest gray and black vertices of the hypercube as robustness measure $\mathcal{R}$. It can be computed efficiently by solving the optimization problem

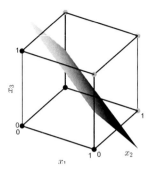

Figure 3.4.: Linear separation of the vertices with $B(x) = 0$ (black) and those with $B(x) = 1$ (gray) of the unate Boolean function $B(x) = (x_1 \text{ and } x_2) \text{ or } (x_3 \text{ and } (x_1 \text{ or } x_2))$.

$$
\begin{aligned}
\max_{\mathcal{R}, a_0 \in \mathbb{R},\ a \in \mathbb{R}^{|J|}} \quad & \mathcal{R} \\
\text{s.t.} \quad & \forall l \in \{1, \ldots, m \mid y_i^l = 1\} : \ a_0 + a^T x_J^l \geq \mathcal{R} \\
& \forall l \in \{1, \ldots, m \mid y_i^l = 0\} : \ a_0 + a^T x_J^l \leq -\mathcal{R} \\
& \|a\|_2 \leq 1.
\end{aligned}
\tag{3.26}
$$

In this, x_J is the subvector of x containing the three potential regulators of x_i. Note that Problem (3.26) is a convex optimization problem with linear cost function and quadratic constraints. Furthermore, Problem (3.25) is contained in it: if $\mathcal{R} \leq 0$, Problem (3.25) is infeasible, otherwise it is feasible.

Roughly speaking, we intend to maximize the value $P(x^l)$ for each x^l for which we have observed a successor $y^l = 1$, and minimize $P(x^l)$ for each x^l for which a successor $y^l = 0$ has been observed. Combinations of regulators that achieve a higher value $\mathcal{R}$ are considered as biologically more plausible. While this definition of $\mathcal{R}$ is heuristic in nature, it is shown next with an example that it can indeed be related to robustness properties of the biological mechanisms realizing a given update function. Assume that, for node x_1 of a network, the following two update functions can both explain all observed transition pairs.

(I) $B_1 = B_1(x_2, x_3, x_4)$, and the measurement set $\mathcal{O}$ requires that B_1 satisfies $B_1(1,0,0) = 1$, $B_1(0,1,0) = 1$, $B_1(0,1,1) = 0$ and $B_1(1,1,0) = 0$. The remaining vertices are not specified by $\mathcal{O}$.

(II) $B_1 = B_1(x_5, x_6, x_7)$, and the measurement set $\mathcal{O}$ requires that B_1 satisfies $B_1(0,1,0) = 1$, $B_1(0,0,1) = 1$, $B_1(1,0,0) = 0$ and $B_1(0,1,1) = 0$. The remaining vertices are not specified by $\mathcal{O}$.

First note that both cases can indeed be realized by a unate Boolean function. Figure 3.5 illustrates these two cases. Also the best separating hyperplanes were computed according to (3.26) yielding $\mathcal{R} = 0.29$ for case (I), and $\mathcal{R} = 0.20$ for case (II). This means that the position of the hyperplane, that is, the optimal parameters a_0 and a defining the hyperplane, can be varied more in case (I) without violating the requirements

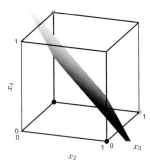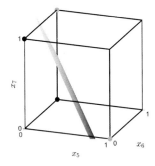

Figure 3.5.: Illustration of case (I) (left) and case (II) (right) and the respective optimal separating hyperplanes.

on the respective update function imposed by the data. Biologically, this allows for the interpretation that the molecular mechanisms responsible to generate the update function need to be tuned finer, and are therefore less robust in case (II). Indeed, in case (I), the optimal update function resulting from Problem 3.26 assigns the following Boolean values to the vertices not defined by the data: $B_1(0,0,0) = 1$, $B_1(0,0,1) = 1$, $B_1(1,0,1) = 0$ and $B_1(1,1,1) = 0$. However, this is not the only possibility. It is also possible to choose $B_1(0,0,1) = 0$ instead of $B_1(0,0,1) = 1$, or $B_1(1,0,1) = 1$ instead of $B_1(1,0,1) = 0$, and still obtain a unate function $B_1(x_2,x_3,x_4)$. In case (II), the assignment of truth values to vertices as defined by the optimal hyperplane is the only possible way to obtain a unate function $B_1(x_5,x_6,x_7)$. Thus, there are more degrees of freedom to realize the update function in case (I) than in case (II).

Unfortunately, when considering update functions of at most three arguments, there are not many cases in which different possible update functions can achieve different values of $\mathcal{R}$ according to (3.26). Indeed, every unate Boolean update function of two arguments that cannot be reduced to a Boolean function of less arguments achieves the same value of $\mathcal{R}$. In the case of three arguments, only cases which are symmetric to (I) and (II) can be distinguished by $\mathcal{R}$. In higher dimensions however, the measure will become more relevant, although not all unate Boolean functions are captured any longer by the sign-representation using affine polynomials. In the Boolean application example in the next section, the advantage of the very simply optimization problem (3.26) compared to the more complicated formulations derived in the last section will become obvious.

3.5. Application examples

In order to illustrate the benefits of the linear reformulation of the problems to find optimal update functions developed in Sections 3.3 and 3.4, we now apply these methods to reconstruct two models from artificial data sets. The first example deals with the reconstruction of a Boolean model using synchronous updates. It will be discussed in greater detail than the second example, which considers the discrete setting and asynchronous updates and serves mainly as a proof of principle. All numerical experiments were executed on a common laptop computer (i5 520M CPU, 4GB RAM)) using the CPLEX solvers contained in the IBM ILOG suite via the Matlab interface. As an alternative collection of linear and mixed integer linear programming solvers, the GNU linear programming kit (GLPK) was used, too.

3.5.1. A Boolean example

Albert & Othmer (2003) and Chaves *et al.* (2005) have presented a Boolean model which describes the regulatory relations between the segment polarity genes in Drosophila melanogaster. This group of genes is involved in the process of forming and maintaining the segments of the body of a fruit fly. The equations shown in Figure 3.6 represent a segment of four cells that repeats periodically. In this, a function B_{x_i} is the update rule for node x in cell $i \in \{1, \ldots, 4\}$. The model thus contains 52 variables. A detailed description and analysis can be found in Albert & Othmer (2003) and Chaves *et al.* (2005).

From this model, a data set $\mathcal{O}$ was generated, which comprises the model's six steady states, half of which have biologically observed phenotypes, and one trajectory, starting at a well-defined wild-type initial condition and reaching the wild-type steady state in six update steps. The set $\mathcal{O}$ thus contains 12 transition pairs, 6 for the steady states and 6 for the wild-type trajectory. In the following, several problems concerning the identification of a Boolean model from this data set are studied.

Identification of the interaction structure

The first problem is to compute all minimal sets of activators and inhibitors for every node x, which can explain all observed transition pairs in $\mathcal{O}$. Thereby, it is biologically reasonable to assume that, for each secreted factor x, either both or none of the variables x_{i-1} and x_{i+1} enter an update function B. Moreover, their effect on the output of B should be the same. This can be achieved by additional equality constraints on the parameter vector a of P representing B as follows: index the k arguments of P with $j_1, j_2, \ldots, j_k$, and let j_1 and j_2 index x_{i-1} and x_{i+1}, respectively. Then it has to hold that

$$\forall J \subseteq \{j_3, \ldots, j_k\} : a_{\{j_1, J\}} = a_{\{j_2, J\}}.$$

One can verify that these constraints do not affect the applicability of Corollary 3.16. With this, Algorithm 3.1 is executed for each node of the system. Thereby, all nodes from cell i and the secreted factors WG, HH (and also hh) from the two neighboring cells $i + 1$ and $i - 1$ are considered as possible regulators for a node x in cell i. The most outer loop of the algorithm is not continued for $j + 1$ regulators, if the optimal cost function value 0 has been obtained for some combination of j regulators.

While the linear reformulation does not allow to extract more information about the interaction structure from $\mathcal{O}$ than any other formulation, it is computationally efficient. The required CPU time to solve all the linear programs during the execution of Algorithm 3.1 was approximately 15 seconds, the solution of all mixed integer linear programs took approximately 40 seconds. The minimal sets of regulators for all nodes except CIA and CIR are summarized in Table 3.1. One can see that for nodes WG, en, EN, HH, PTC, ci, and CI, all regulators could be correctly identified as activators or inhibitors. For wg, the data set did not contain sufficient information to derive all

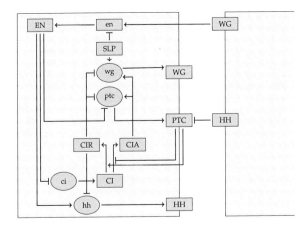

$$B_{\mathrm{SLP}_i} = \begin{cases} 0 & \text{for } i = 1,2 \\ 1 & \text{for } i = 3,4 \end{cases}$$

$$
\begin{aligned}
B_{\mathrm{WG}_i} &= wg_i & B_{\mathrm{en}_i} &= (\mathrm{WG}_{i-1} \vee \mathrm{WG}_{i+1}) \wedge \neg\mathrm{SLP}_i \\
B_{\mathrm{EN}_i} &= en_i & B_{\mathrm{hh}_i} &= \mathrm{EN}_i \wedge \neg\mathrm{CIR}_i \\
B_{\mathrm{HH}_i} &= hh_i & B_{\mathrm{ptc}_i} &= \mathrm{CIA}_i \wedge \neg\mathrm{EN}_i \wedge \neg\mathrm{CIR}_i \\
B_{\mathrm{ci}_i} &= \neg\mathrm{EN}_i & B_{\mathrm{CI}_i} &= ci_i \\
B_{\mathrm{PTC}_i} &= ptc_i \vee (\mathrm{PTC}_i \wedge \neg\mathrm{HH}_{i-1} \wedge \neg\mathrm{HH}_{i+1}) \\
B_{\mathrm{wg}_i} &= (\mathrm{CIA}_i \wedge \mathrm{SLP}_i \wedge \neg\mathrm{CIR}_i) \vee (wg_i \wedge (\mathrm{CIA}_i \vee \mathrm{SLP}_i) \wedge \neg\mathrm{CIR}_i) \\
B_{\mathrm{CIA}_i} &= \mathrm{CI}_i \wedge (\neg\mathrm{PTC}_i \vee \mathrm{HH}_{i-1} \vee \mathrm{HH}_{i+1} \vee hh_{i-1} \vee hh_{i+1}) \\
B_{\mathrm{CIR}_i} &= \mathrm{CI}_i \wedge \mathrm{PTC}_i \wedge \neg\mathrm{HH}_{i-1} \wedge \neg\mathrm{HH}_{i+1} \wedge \neg hh_{i-1} \wedge \neg hh_{i+1}
\end{aligned}
$$

Figure 3.6.: Schematic illustration and update rules of the Boolean model of the segment regulatory genes. Pointed edges denote activating influences, blunt edges denote inhibiting interactions. The symbols $\neg$, $\wedge$, $\vee$ denote the logic operators not, and, and or, respectively. The figure and the equations are adopted from Chaves *et al.* (2005).

regulators, but it could be identified to be self-activating. For the two nodes with the largest number of regulators, CIA and CIR, there were many (11 for CIA and 14 for CIR, see Table 3.2) minimal sets of regulators, with the true combination being only one of them. In the following, the nodes CIR and CIA are therefore examined in more detail.

In depth analysis of nodes CIA and CIR

We investigate the question whether the recorded transitions of CIR and CIA can be explained without intercellular communication, that is, excluding $\mathrm{WG}_{i\pm1}$, $hh_{i\pm1}$, and $\mathrm{HH}_{i\pm1}$ as regulators. As a result of the optimization in which we have searched for arbitrary, and not only unate functions, the existence of such an update function could be excluded, thus confirming the need for intracellular communication. The CPLEX linear programming solver needed less than one second to give this answer.

Table 3.1.: Minimal sets of regulators. Activators are marked with "+", inhibitors with "−".

node	regulators
wg_i	$\{wg^+\}$
WG_i	$\{wg_i^+\}$
en_i	$\{SLP_i^-, WG_{i\pm1}^+\}$
EN_i	$\{en_i^+\}$
hh_i	$\{EN_i^+\}$
HH_i	$\{hh_i^+\}$
ptc_i	$\{en_i^-, CIA_i^+\}, \{EN_i^-, CIA_i^+\}, \{HH_i^-, CIA_i^+\}, \{CI_i^+, CIA_i^+\}$
PTC_i	$\{ptc_i^+, PTC_i^+, HH_{i\pm1}^-\}$
ci_i	$\{EN_i^-\}$
CI_i	$\{ci_i^+\}$

A second question addresses a particularity of the model as presented in Figure 3.6. CIA and CIR are regulated by $hh_{i\pm1}$, which represent mRNA concentrations. However, mRNAs cannot, in general, directly influence the expression of genes, which makes the presence of these variables in the update functions B_{CIA} and B_{CIR} artificial. We therefore ask the question, if the observed transitions of CIA and CIR can be explained by a combination of regulators among proteins only. To this end, Algorithm 3.1 is applied to find potential unate update functions. The result shows that such update functions indeed exist. All possible minimal sets of regulators are listed in Table 3.2.

In fact, the presence of $hh_{i\pm1}$ in the update rules B_{CIR} and B_{CIA} is an uncertain point in the model (see the Appendix in Albert & Othmer (2003)). While the formations of CIA and CIR from CI involve various other protein complexes, the mechanisms are not yet completely understood (Ogden *et al.*, 2004). Our analysis suggests an alternative way of modeling the formation of these proteins, replacing $hh_{\pm1}$ by CIA as regulator of CIA and CIR.

Especially for this second question, the advantage of the linear formulation becomes obvious. The CPLEX simplex solver needed approximately 1000 seconds, while the CPLEX mixed integer linear programming solver needed approximately 3700 seconds. Using the GLPK solvers, the difference was even more pronounced. While the simplex solver terminated in approximately 2000 seconds, the mixed integer solver did not terminate within 8 days.

Identification using the reduced order representation

In a next step, the identification process is repeated using the more compact formulation in Problem (3.25) which is based on the sign-representation of unate Boolean functions by affine polynomials. Interestingly, the optimization yields exactly the same results as the analysis before, which was based on the more complex representation by multi-affine polynomials (see Table 3.1). However, the computing time was reduced to approximately 5 seconds, as only linear feasibility problems of complexity linear in the problem dimension need to be solved. These results thus suggest to use the sign-representation by simple affine polynomials instead of the multi-affine representation as in Equation (3.11) also for large systems, as the significant reduction of the memory requirements and the computing time justifies the loss of generality.

Additionally, the robustness scores according to (3.26) have been computed for all possible sets of regulators. Unfortunately, the minimal sets of regulators for ptc could not be distinguished by this measure, as all sets lead all to a robustness value of $\mathcal{R} = 0.35$. For the possible sets of regulators for nodes CIA and CIR, the sets containing $hh_{\pm 1}$ were among the sets with the smallest value for $\mathcal{R}$, supporting the above argumentation that this peculiarity of the models is indeed questionable. Our alternative suggestion to replace $hh_{\pm 1}$ by CIA yields a higher robustness score (see Table 3.2).

Table 3.2.: Minimal sets of regulators for nodes CIR and CIA together with the achieved robustness value. Activators are marked with "+", inhibitors with "−".

regulators for CIA_i	$\mathcal{R}$	regulators for CIR_i	$\mathcal{R}$
$\{SLP_i^-, wg_i^+, CI_i^+, CIR_i^-\}$	0.25	$\{SLP_i^+, wg_i^-, CI_i^+, CIR_i^+\}$	0.25
$\{SLP_i^-, WG_i^+, CI_i^+, CIR_i^-\}$	0.25	$\{SLP_i^+, WG_i^-, CI_i^+, CIR_i^+\}$	0.25
$\{HH_{i\pm1}^+, ptc_i^+, PTC_i^-, CI_i^+\}$	0.18	$\{HH_{i\pm1}^-, ptc_i^-, PTC_i^+, CI_i^+\}$	0.22
$\{HH_{i\pm1}^+, PTC_i^-, CI_i^+, CIA_i^+\}$	0.18	$\{en_i^-, HH_{i\pm1}^-, CI_i^+, CIA_i^-\}$	0.22
$\{HH_{i\pm1}^+, ci_i^-, CI_i^+, CIA_i^+\}$	0.18	$\{hh_i^-, HH_{i\pm1}^-, CI_i^+, CIA_i^-\}$	0.22
$\{en_i^+, HH_{i\pm1}^+, CI_i^+, CIA_i^+\}$	0.18	$\{HH_{i\pm1}^-, PTC_i^+, CI_i^+, CIA_i^-\}$	0.22
$\{hh_i^+, HH_{i\pm1}^+, CI_i^+, CIA_i^+\}$	0.18	$\{HH_{i\pm1}^-, ptc_i^-, PTC_i^+, ci_i^+\}$	0.22
$\{en_i^+, hh_{i\pm1}^+, HH_{i\pm1}^+, CI_i^+\}$	0.13	$\{HH_{i\pm1}^-, PTC_i^+, ci_i^+, CIA_i^-\}$	0.22
$\{hh_i^+, hh_{i\pm1}^+, HH_{i\pm1}^+, CI^+\}$	0.13	$\{HH_{i\pm1}^-, ci_i^+, CI_i^+, CIA_i^-\}$	0.22
$\{hh_{i\pm1}^+, HH_{i\pm1}^+, PTC_i^-, CI_i^+\}$	0.13	$\{hh_{i\pm1}^-, HH_{i\pm1}^-, PTC_i^+, ci_i^+\}$	0.20
$\{hh_{i\pm1}^+, HH_{i\pm1}^+, ci_i^-, CI_i^+\}$	0.13	$\{en_i^-, hh_{i\pm1}^-, HH_{i\pm1}^-, CI_i^+\}$	0.20
		$\{hh_{i\pm1}^-, HH_{i\pm1}^-, PTC_i^+, CI_i^+\}$	0.20
		$\{hh_i^-, hh_{i\pm1}^-, HH_{i\pm1}^-, CI_i^+\}$	0.20
		$\{hh_{i\pm1}^-, HH_{i\pm1}^-, ci_i^+, CI_i^+\}$	0.20

3.5.2. A discrete example

While the previous example dealt only with the Boolean case, we now turn to the more general and complex class of discrete systems. The model we consider here was originally developed by Ropers *et al.* (2006) as a system of piecewise affine ordinary differential equations. It describes the response of a gene regulation network of Escherichia coli cells to carbon deprivation, which causes the cells to leave the so-called exponential-growth rate and enter a more resistant and energy saving stationary phase. This model was converted into a discrete model by Chaves *et al.* (2010). It involves six genes, crp, cya, gyrAB, topA, fis and rrn. The carbon starvation signal is denoted by u. As the update functions of the discrete model are too space consuming, they are not repeated here and we refer to Chaves *et al.* (2010) instead. Figure 3.7 shows an illustration of the mutual influences between the genes in this network.

The variables have ranges crp, cya, gyrAB, topA $\in \{0, 1, 2\}$, fis $\in \{0, 1, 2, 3, 4\}$ and u $\in \{0, 1\}$. The discrete model exhibits two attractors. The first attractor for $u = 1$,

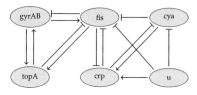

Figure 3.7.: Interaction structure of the carbon starvation network. Pointed edges denote activating influences, blunt edges denote inhibiting interactions.

denoted by $\mathcal{A}_1$, is characterized by crp $= 2$, cya, gyrAB ≥ 1 and topA $=$ fis $= 0$. The second attractor for $u = 0$, denoted by $\mathcal{A}_2$, is determined by crp $= 1$, cya $= 2$, topA ≤ 1, and fis ≥ 1. Setting $u = 1$ ($u = 0$), the discrete system will finally always reach attractor $\mathcal{A}_1$ ($\mathcal{A}_0$).

From this model, an artificial measurement set $\mathcal{O}$ was recorded by letting the system start at random positions in attractor $\mathcal{A}_1$ ($\mathcal{A}_0$) and stimulating the system with $u = 0$ ($u = 1$) until the other attractor was reached. Altogether, 15 trajectories from $\mathcal{A}_0$ to $\mathcal{A}_1$, and 15 trajectories from $\mathcal{A}_1$ to $\mathcal{A}_0$ were recorded from simulations. Thereby, asynchronous updates were used, which also implicates that not all transition pairs in $\mathcal{O}$ can be used for the estimation of an update function B_i. As outlined in Section 3.1.1, only those transition pairs (x^l, y^l) were considered for which it holds that $y_i^l \neq x_i^l$. The knowledge of the asymptotic behavior of the network was included into the identification procedure, too. Assume we want to find an update function $B_i(x_J)$. Then, all states x_J in attractor $\mathcal{A}_0$ ($\mathcal{A}_1$) are enumerated and it is required that $B_i(x_J)$ is larger or smaller than the smallest or respectively largest value of x_i in attractor $\mathcal{A}_0$ ($\mathcal{A}_1$). This leads to linear constraints of the form $ae(x_J) \leq (\geq)b$. Considering the unateness properties, we only require that each B_i is unate in all arguments x_j, $j \neq i$. The reason why x_i is excluded is the self-degradation of x_i. It might for example be possible that x_i is self-activating such that B_i is increasing in x_i for small values of x_i, but decreasing in x_i for large values. Note however, that this distinction has no effect in the Boolean case.

With these modeling assumptions, Algorithm 3.1 is applied. The search for all possible combinations of up to all 6 regulators and all assignments of activators or inhibitors took approximately 90 minutes. Thereby, the linear program had to be solved 1305 times. The identified minimal sets of regulators that can explain the observed transitions and guarantee that the sets $\mathcal{A}_0$ and $\mathcal{A}_1$ are forward-invariant, are listed in Table 3.3.

For all nodes except topA, the listed minimal sets of regulators were contained in all other possible sets of regulators. For topA, also sets containing $\{crp, topA, u\}$ and arbitrary other regulators are possible. The correct interactions were always among the possible solutions and the major part of regulators could be identified correctly. Thus, the example serves as a proof of principle and shows that the proposed method can also be applied to the rather difficult problem of identifying discrete models under asynchronous updates.

Table 3.3.: Minimal sets of regulators and their roles. Uniquely identified activators are marked by "$+$", inhibitors by "$-$".

Node	minimal sets of regulators
crp	$\{\mathrm{crp,fis^-}\}, \{\mathrm{crp,u^+}\}, \{\mathrm{fis^-,u}\}, \{\mathrm{topA^+,u}\}$
cya	$\{\mathrm{cya,u^-}\}$
fis	$\{\mathrm{fis,gyrAB^+,u}\}$
gyrAB	$\{\mathrm{fis^-,gyrAB}\}$
topA	$\{\mathrm{fis^+,topA}\}$

3.6. Summary and discussion

In this chapter, the problem of identifying discrete models of gene regulation networks from discretized and error prone measurements was studied. This problem was first formulated as a combinatorial optimization problem. We have developed a novel polynomial representation of Boolean and discrete functions, which has then been employed to reformulate the combinatorial as a linear optimization problem. This has also lead to a complete algorithm for the identification problem.

For the class of unate Boolean functions, we have furthermore developed a more compact sign-representation by affine polynomials. Although giving up on generality, the sign-representation allows for significantly reduced optimization problems compared to the general case. For the same class of functions, we have furthermore developed a novel robustness measure for the robust reproduction of observed trajectories. The measure can be computed by a quadratically constraint optimization problem.

The advantage of the reformulation of the identification problem as a linear program was finally demonstrated in a Boolean and a discrete case study. As a result of this chapter, we now have a very general formulation that can be used to identify arbitrary Boolean or discrete functions, and a very efficient formulation to only search for unate Boolean functions.

Let us summarize this chapter again by referring to the goals formulated in Section 3.1.3. We have achieved several formulations of the identification problem as linear programs, which allows the application of highly optimized linear programming solvers. Also the desired flexibility is given. No new optimization algorithms need to be designed if the desired properties of the estimated update functions change. It is sufficient to simply add constraints to the optimization problem. This has been shown for the unateness and the canalizing property, but also for the requirement that two arguments should enter an update function in the same way in the Boolean example. As the optimization problems return a complete Boolean or discrete function, the result can be used for simulating the system behavior. Also the last goal to incorporate robustness considerations into the identification procedure has been addressed for the relevant class of unate Boolean functions.

4. Analysis of multistability and multistability robustness

In this chapter the problems of model validation and steady state robustness analysis with respect to a desired multistable behavior are studied. Section 4.1 introduces these problems in detail, discusses related approaches in the literature and summarizes the goals and challenges for this chapter. In Section 4.2, a qualitative modeling framework for gene regulation networks based on differential equations and some mathematical preliminaries are presented. In Chapter 4.3, a first method for model validation is developed. A second and refined method for model validation and robustness analysis is derived in Section 4.4. The benefits of these two methods are demonstrated with the help of two application examples in Section 4.5. The chapter concludes with a summary and discussion in Section 4.6.

The methods and results presented in this chapter have been published previously in Breindl & Allgöwer (2009); Breindl et al. (2010, 2011a,b).

4.1. Introduction and problem statement

Multistability is a recurrent phenomenon for biological systems, and for gene regulation networks in particular. In this context, it is an broadly accepted hypothesis that multistability is essential for cell differentiation, such that each cell type can be characterized by a stable expression pattern of the underlying gene regulation network. This idea dates back to the early works of Kauffman (Kauffman, 1969, 1993) and has been confirmed by more recent studies (Aldana *et al.*, 2007), prominent examples for which are hematopoietic or mesenchymal stem cells (Roeder & Glauche, 2006; Lemon *et al.*, 2007; Krumsiek *et al.*, 2011). Pattern formation is another field that is often related to multistable gene regulation networks. The Drosophila polarity gene network which has already been studied in the last chapter is one example, the network underlying the formation and maintenance of the midbrain-/hindbrain boundary in mammals (Wittmann *et al.*, 2009b) is another one. In these examples, the spatial patterns are generated by cells, in which a certain gene regulation network operates at different steady states. Further examples for multistable networks are the lactose utilization network of Escherichia coli (Ozbudak *et al.*, 2004), which regulates the expression of the lac operon depending on the presence of glucose and lactose, or the apoptosis signaling network (Wajant *et al.*, 2003; Eissing *et al.*, 2004), which acts as a switch between death or survival of a cell.

In this chapter we consider multistable gene regulation networks, and assume that several steady states of such a network have been observed experimentally. We assume furthermore, that there is an hypothesis about the interaction structure of the network, while the detailed kinetics are largely unknown. This is a reasonable assumption, as knockout or over-expression experiments and the application of methods as developed in Chapter 3 allow for qualitative conclusions about the interaction structure, such as the existence and the sign of an interaction between two genes. The typically large errors of common measurement techniques however often prevent the identification of the exact kinetics. For this setup, we study the important questions of model validation an discrimination, as introduced next.

4.1.1. Problem formulation

For the scenario described above, a very basic validation problem is formulated. It amounts to ask, if the hypothesis about the interaction structure can explain the observed multistable behavior, or if it can be rejected based on the observations.

For the problem of model discrimination, consider the case, that there exist several hypotheses which cannot be rejected. They may for example differ in the roles of some proteins, such that protein A inhibits the expression of some other protein B according to one hypothesis, but activates it according to another. The study of a gene regulation network maintaining a well-defined expression pattern at the mid-hindbrain boundary (Wittmann *et al.*, 2009b) is an example for such a situation. It is then a relevant problem to find additional criteria to compare the different hypotheses. A good basis for this is the concept of robustness, which is defined according to Kitano (2004) as a system's ability to maintain its function in the presence of perturbations. Assuming that biological systems have evolved in a way that has made them robust against common perturbations, we consider the most robust system, that is, the system that can tolerate the largest perturbations until it loses its function, as biologically most plausible. In the context of this chapter, and as mentioned before, the function that shall be realized by the network is an observed multistable behavior. Of course, depending on the concrete networks to be studied, also other network functions can be reasonable. This is however not in the scope of this thesis.

Finally, these problems need to be formulated in a mathematical framework, which can address the uncertainty about the biological knowledge and the data in an appropriate way, and which provides the necessary tools to arrive at efficient algorithmic solutions. We next review several frameworks and approaches toward these problems. Based on this, a specific framework will be selected in Section 4.1.3, and the problem descriptions will be made more precise.

4.1.2. Established approaches

The problem to compute a system's steady states and to determine their stability properties has been a fundamental topic in the systems theory literature ever since and, for systems of ordinary differential equations, whole text books are devoted to it. Their review is however not in the scope of this chapter, and only a few nonstandard approaches shall be mentioned which have been developed in the context of systems biology and take the specific structure of these systems into account.

Using monotone systems theory, Angeli & Sontag (2003, 2004) have shown that all steady states of a monotone system and their stability properties can be determined graphically by a certain input-output characteristic. While this concept is very appealing, it requires a model with exactly known parameters or regulatory functions, which is usually not available for uncertain gene regulation networks.

Therefore, also frameworks have been proposed that put more emphasis on the interaction structure rather than on a specific choice of parameters. Boolean or discrete systems (Kauffman, 1969; Thomas, 1973; Klamt *et al.*, 2006) are the most important frameworks to be mentioned in this context and they have already been studied under the aspect of structure estimation in the last chapter. The problem of finding all attractors of these systems has been studied extensively in the literature. Zhang *et al.* (2007) for example present algorithms to identify all singleton and small attractors of Boolean networks, and show that this problem is NP-hard.

A further framework which still puts more emphasis on the interaction structure than on a specific choice of parameters is the class of piecewise linear models. For this class, only the relative ordering of several threshold parameters has to be specified. The attractors of such a system have been studied in several publications (Snoussi, 1989; de Jong *et al.*, 2004; Casey *et al.*, 2006; Batt *et al.*, 2008) using methods from both systems theory and computer science.

The problem of investigating the robustness properties and, in particular, the robustness of a system's steady states against perturbations, has a long tradition in systems theory, too. Again, we do not intend to give an overview of the rich literature on robustness and sensitivity analysis, but only mention some relevant approaches specifically developed for biological networks and their steady state behavior.

Let us start with an overview of approaches based on a Boolean or discrete description. In this context, a frequently studied question is that of how the attractor landscape is affected by perturbations of the network. Aldana *et al.* (2007) for example, have performed an extensive numerical study, in which thousands of randomly generated Kauffman networks (Kauffman, 1969) are modified by duplicating one gene and modifying some update functions related to that gene. This is is a widely accepted hypothesis of how networks might evolve. The authors have found that networks which operate close to the critical regime (see Derrida & Stauffer (1986)) maintain most old but may exhibit some additional attractors after that modification, making these networks robust but evolvable. Balleza *et al.* (2008) have confirmed with the help of gene expression data of real networks, that these systems indeed operate close to criticality. Besides studying the effects of such structural changes of the attractor landscape, other approaches investigate the effects of randomizing the order of updating the individual nodes. Chaves *et al.* (2005) for example have shown that the biologically relevant attractors of the Drosophila polarity gene network, which was introduced in the last chapter, are robustly maintained under this type of perturbations.

For biological systems described by ordinary differential equations, several approaches to analyze the robustness of steady states with respect to perturbations have been presented, too. Eißing *et al.* (2005) for example study the question of how robustly the bistable behavior of an apoptosis signaling network is maintained with respect to parametric perturbations or noise. First order sensitivity analysis as well as statistical methods are used for this purpose. Jacobsen & Cedersund (2008) have introduced so-called dynamic perturbations which represent uncertainties about the interactions or unmodeled dynamics as dynamic systems. The robustness of the model against these uncertainties is then analyzed using methods from robust control theory, which rely on a linearization of the original system. A large variety of methods which allow a more global view on the dependency of the steady states on parameters, and do not require a linear or linearized model, are presented by Waldherr *et al.* (2008), Waldherr *et al.* (2009), Waldherr & Allgöwer (2011), and Waldherr *et al.* (2011). These methods make use of systems and control theoretical concepts such as the Nyquist criterion and results from convex optimization, such as semi-definite programming relaxations of polynomial optimization problems (Parrilo, 2003) or the Positivstellensatz. These methods allow for conclusions about the existence or the stability of steady states for a wide range of parameter values. However, all approaches have in common that they require a model description with well-defined parameter values or ranges.

While the works cited above are motivated mainly from a theoretical perspective, there is also a large number of studies which focus on a specific model and rely on sampling and simulation techniques in order to arrive at conclusions about the system's robustness or sensitivity. Two examples that again focus on the Drosophila segmentation gene network are the case studies by von Dassow *et al.* (2000) and Ingolia (2004), which show that the polarity segmentation regulatory network in Drosophila indeed robustly maintains its desired multistability properties against large ranges of parameter variations.

Finally, a very interesting approach toward a steady state robustness analysis which shares several similarities with the work in this chapter has been presented by Blanchini & Franco (2011). Lyapunov and invariant set theory are employed to show that a system of ordinary differential equations exhibits one or several equilibria as long as the interaction functions satisfy some structural properties, such as for example a sigmoidal form. The stability properties are therefore robust against all perturbations that do not change these fundamental structural properties.

4.1.3. Challenges and goals

All modeling and analysis frameworks presented in the last section have disadvantages which prevent efficient solutions to our validation and discrimination problem. For a Boolean or discrete framework, several ways to define the robustness of a system's steady state behavior have been suggested, but the computation of a robustness measure will always have to resort to extensive simulations. For a framework based on ordinary differential equations, all but the last approach described in Section 4.1.2 require a fully parameterized system and precise steady state locations. The work by Blanchini & Franco (2011) is most likely not accessible to an algorithmic solution. We therefore aim to develop a modeling and analysis framework which combines the advantage of discrete or Boolean systems, that no detailed knowledge about the system's kinetics and steady state locations is necessary, with the advantage of a description by differential equations which allows access to algorithmic tools from systems and control theory. The individual goals for this chapter can now be made more precise.

Similar to a Boolean description, the modeling approach should emphasize the interaction structure of the system more than kinetic details, that is, it should not be necessary to make specific assumptions about the reaction kinetics or parameter values. The framework should however allow to specify the reaction kinetics and measurements in greater detail than this is possible in a discrete or Boolean framework, if this information is available. The framework should thus exhibit both, continuous and discrete aspects to incorporate qualitative "on / off" statements from the Boolean world as well as continuous numeric values. As we aim to build on a system description based on ordinary differential equations, the system will have the form as introduced in Equation (2.3), that is,

$$\dot{x}_i = f_i(x) = -\gamma_i x_i + r_i(x), \quad i = 1, \ldots, n, \tag{4.1}$$

in which we assume a linear degradation rate for all nodes i in the network. Similar to the work by Blanchini & Franco (2011), we want to be able to specify a hypothesis about the interaction structure by general properties of the functions f_i, while their exact shapes may remain uncertain. A good starting point for such a description, which already exhibits these desired properties to a large degree, has been presented by Chaves *et al.* (2008) and will be further refined and formalized in this chapter. The methods for model validation and discrimination will then build on this framework.

For the model validation problem, the goal is to find a fast and efficient method, which can approve or reject a hypothetical model only on the basis of the general properties of the right-hand side functions. For the discrimination problem, we require the definition of a plausibility measure, which is based on biological robustness considerations. Using this measure, it should be possible to attach a numeric value to each hypothetical model and thus, to make alternative models comparable. This measure should reflect how well the uncertain model is suited to generate the desired multistable behavior. Furthermore, an efficient formulation to compute this numeric value needs to be found. The main goals of this chapter can thus be summarized as follows.

- Elaboration of a modeling framework based on ordinary differential equations that allows the description of an uncertain system and of uncertain measurements.

- Development of a fast and easily applicable method to test if the uncertain system description can in principle reproduce an observed multistable behavior.

- Definition of a robustness measure that characterizes the system's ability to maintain a certain multistable behavior under perturbations. This also requires means to quantify relevant perturbations.

- Development of a computationally efficient formalism to compute the robustness measure and thus, to be able to compare different hypothetical models.

4.2. Modeling framework and preliminaries

As outlined in the previous sections, we aim to base our modeling framework on only few assumptions in order to allow a large uncertainty in the right-hand side functions $f_i(x)$. A few generally accepted assumption however need to be made.

The first assumption has already been used in Chapter 3.1, and has lead to the definition of unate discrete functions. It requires that transcription factors will always have either an inhibiting or an activating effect on some fixed target gene in the network. In our continuous framework, this assumption leads to the definitions of activation and inhibition functions, which are now recalled from Chaves *et al.* (2008).

Definition 4.1 (Activation function)**.** *Let $M \in \mathbb{R}_+$. An* activation function *is a function* $v : [0, \infty) \to [0, M)$ *with:*

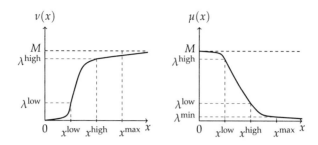

Figure 4.1.: Illustration of an activation and an inhibition function and of the respective critical concentrations and values.

i) v *is continuous,*

ii) $v(0) = 0$ *and* $v(x) \to M$ *as* $x \to \infty$,

iii) $v(x)$ *is monotonically increasing.*

Definition 4.2 (Inhibition function). *Let* $M \in \mathbb{R}^+$. *An inhibition function is a function* $\mu : [0, \infty) \to (0, M]$ *with:*

i) μ *is continuous,*

ii) $\mu(0) = M$ *and* $\mu(x) \to 0$ *as* $x \to \infty$,

iii) $\mu(x)$ *is monotonically decreasing.*

We furthermore denote the set of all activation functions by $\mathcal{N}$, and the set of all inhibition functions by $\mathcal{M}$. The symbol φ is then used to represent either an activation function $v \in \mathcal{N}$, or an inhibition function $\mu \in \mathcal{M}$. Finally, $\mathcal{S}_\varphi$ is the class of a function φ, that is, $\mathcal{S}_\varphi = \mathcal{N}$ if φ is an activation function, and $\mathcal{S}_\varphi = \mathcal{M}$ if φ is an inhibition function. For the argument of an activation or inhibition function $\varphi(x)$, we moreover introduce three values x^{low}, x^{high} and x^{max}, with $0 < x^{\text{low}} < x^{\text{high}} < x^{\text{max}}$, to which we will refer to as *critical concentrations*. For an activation function $v(x)$, we then define $\lambda_v^{\text{low}} := v(x^{\text{low}})$ and $\lambda_v^{\text{high}} := v(x^{\text{high}})$. Furthermore, $\lambda_\mu^{\text{high}} := \mu(x^{\text{low}})$, $\lambda_\mu^{\text{low}} := v(x^{\text{high}})$, and $\lambda_\mu^{\text{min}} := \mu(x^{\text{max}})$ for an inhibition function $\mu(x)$. To these values we will refer to as *critical values*. These definitions are illustrated in Figure 4.1. Also note that the most commonly used reaction kinetics such as Michaelis-Menten or Hill-type kinetics (see Chapter 2.2.2) are covered by this modeling assumption.

As the expression of a gene can be modified by several transcription factors, the second model assumption describes how more complicated right-hand side terms can be constructed. In this work, we consider arbitrary sums and products of individual activation and inhibition functions. In order to achieve a compact notation for the production term $r_i(x)$, we use the symbol "$\circ$" for sums "$+$" as well as for multiplications "$\cdot$". With this, the general form of a production term $r_i(x)$ can be written as

$$r_i(x) = \varphi_{i,1}(x_{j_1}) \circ \ldots \circ \varphi_{i,q_i}(x_{j_{q_i}}), \qquad (4.2)$$

with indices $j_k \in \{1, \ldots, n\}$, $k = 1, \ldots, q_i$, and $q_i \in \mathbb{N}_+$. We will call such a production term a *generalized activation and inhibition function*. To illustrate this notation, consider a function $\varphi_{i,k}(x_{j_k})$. In this, the index i denotes the species which is regulated, the index k enumerates its regulators, and the index j_k specifies the species which represents the k-th regulator. Finally, q_i is the number of regulators influencing the production of x_i. Note that self-regulation is allowed.

Without detailed knowledge of the reaction kinetics, the degradation rates γ_i, the exact shapes of the monotonic functions, and, in particular, their critical values cannot be specified. This leads to an uncertain model, in which only the interaction structure is fixed by the structure of the right-hand side as defined in Equations (4.1) and (4.2). With this, the first goal of having an ODE description that allows a high degree of uncertainty has been achieved. In the next section, we show how uncertain measurements can be represented in this framework and relate them to the definitions given in this section.

Measurements

Because of the importance of multistability for many gene regulation networks as outlined in Section 4.1, we want to focus on measurements recorded at steady states. In this context however, we suggest to treat a steady state of such a network not as a single point in the state space, but to consider a whole region, which should be invariant under the dynamics of the model. This point of view is supported by the facts that natural differences between individual cells render the determination of one exact value for each steady state impossible, and that often large uncertainties are introduced by the measurement technique itself.

We therefore weaken the requirement for exact steady state concentrations and consider concentration intervals. As a further simplification, and again supported by the typically large measurement uncertainties, we only distinguish between high and low concentrations. Therefore, for a species x_i, only intervals of low concentrations $\mathcal{I}^{x_i,l} := [0, x_i^{\text{low}}]$ and intervals of high concentrations $\mathcal{I}^{x_i,h} := [x_i^{\text{high}}, x_i^{\text{max}}]$ are considered, in which the values x_i^{low}, x_i^{high}, x_i^{max} are the critical concentrations, with x_i^{max} being the maximal concentration of species x_i. Naturally, we suggest to consider the hyper-rectangular set $\mathcal{X} := [0, x_1^{\text{max}}] \times \ldots \times [0, x_n^{\text{max}}]$ as forward-invariant, that is, $\forall x_0 \in \mathcal{X}, \forall t > 0 : x(t; x_0) \in \mathcal{X}$, in which $x(t; x_0)$ denotes the solution of (4.1) for the initial condition x_0.

Assume furthermore that m different steady states have been observed experimentally, in which each species can be classified to be present at a low or at a high concentration. To represent these steady states, the hyper-rectangular sets $\mathcal{F}_z = \mathcal{I}_z^{x_1} \times \ldots \times \mathcal{I}_z^{x_n}$, $z = 1, \ldots, m$, are then introduced, in which each interval $\mathcal{I}_z^{x_i}$ is either an interval $\mathcal{I}_z^{x_i,l}$, or a interval $\mathcal{I}_z^{x_i,h}$. As a further natural requirement, all sets $\mathcal{F}_z$ should be forward-invariant under the dynamics of the model. In the case that no numeric values can be assigned to the variables x_i^{low}, x_i^{high}, and x_i^{max}, we obtain an uncertain description for the observed multistable behavior.

Remark 4.3. *Depending on the observed steady states, not always all critical concentrations x_i^{low}, x_i^{high}, and x_i^{max} for all species x_i are needed to describe the forward invariant sets. For notational simplicity we will however always use all of them and omit the distinction of cases.*

Remark 4.4. *For the description of measurements, we have assumed that there is only one set of critical concentrations for each species. This means for example that, if there are two different steady states in both of which x_i is present at high concentrations, then the same interval $\mathcal{I}^{x_i,h}$ has to be used to characterize those two steady states. While the modeling framework can be generalized to an arbitrary number of critical concentrations, this would however unnecessarily complicate the presentation and is therefore omitted.*

Example 4.5. *Let us illustrate the concept of forward-invariant sets composed of low or high intervals for each state variable with the help of the mutual inhibition network*

$$\dot{x}_1 = -\gamma_1 x_1 + \mu_{1,1}(x_2)$$
$$\dot{x}_2 = -\gamma_2 x_2 + \mu_{2,1}(x_1). \tag{4.3}$$

The phase diagram of this system is shown in Figure 4.2 for a specific choice of parameters γ_1 and γ_2 and inhibition functions $\mu_{1,1}$ and $\mu_{2,1}$. The nullclines $\dot{x}_1 = 0$ and $\dot{x}_2 = 0$ are depicted in that figure as well. One can see that this system exhibits three steady states, located at the intersections of the nullclines. The outer two steady states are stable. To describe these two steady states in the framework introduced above, we are not interested in their precise positions. It suffices to specify the two forward-invariant sets $\mathcal{F}_1$ and $\mathcal{F}_2$ that contain the corresponding steady states and can be extended in each variable x_i to either the lower boundary $x_i = 0$ or the upper boundary $x_i^{\max}$ of $\mathcal{X}$.

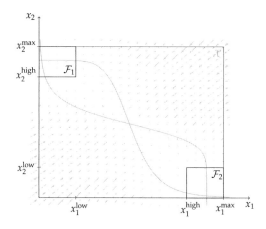

Figure 4.2.: Phase portrait of the mutual inhibition network and forward-invariant-sets $\mathcal{X} = [0, x_1^{\max}] \times [0, x_2^{\max}]$, $\mathcal{F}_1 = [0, x_1^{\text{low}}] \times [x_2^{\text{high}}, x_2^{\max}]$, and $\mathcal{F}_2 = [x_1^{\text{high}}, x_1^{\max}] \times [0, x_2^{\text{low}}]$.

Mathematical preliminaries on forward-invariance

As introduced above, the function which has to be realized by a model of the gene regulation network is the forward-invariance of the sets $\mathcal{X}$ and $\mathcal{F}_z$, $z = 1, \ldots, m$. In this section, we will now derive conditions on the monotonic activation and inhibition functions such that these sets are indeed forward-invariant. This derivation will be based on Nagumo's theorem, which is recalled next from Blanchini (1999).

Theorem 4.6 (Nagumo's theorem). *Consider the system $\dot{x} = f(x)$, $x \in \mathbb{R}^n$. Let $\mathcal{F} \subseteq \mathbb{R}^n$ be a closed and convex set. Then $\mathcal{F}$ is forward-invariant if and only if*

$$\forall x \in \mathcal{F}: \quad f(x) \in \mathcal{K}_{\mathcal{F}}(x), \tag{4.4}$$

where $\mathcal{K}_{\mathcal{F}}(x)$ is the tangent cone to $\mathcal{F}$ in x.

In words, Nagumo's theorem states that, at each point x on the boundary of $\mathcal{F}$, the vector field has to be directed inward or tangent to $\mathcal{F}$. Further details and explanations can be found in Blanchini (1999). As the sets $\mathcal{X}$ and $\mathcal{F}_z$ considered here are hyperrectangular and thus convex, this theorem is directly applicable to our model description according to Equations (4.1) and (4.2). This leads to the following result.

Proposition 4.7. *A hyperrectangular set $\mathcal{F} = \mathcal{I}^{x_1} \times \ldots \times \mathcal{I}^{x_n}$, with $\mathcal{I}^{x_j} = [\underline{x}_j, \overline{x}_j]$, $0 \le \underline{x}_j \le \overline{x}_j$, is forward-invariant under the dynamics of the system if and only if*

$$\forall i \in \{1, \ldots n\}: \quad -\gamma_i \underline{x}_i + \underline{\lambda}_{i,1} \circ \ldots \circ \underline{\lambda}_{i,q_i} \ge 0 \tag{4.5}$$

where $\underline{\lambda}_{i,k} = \min_{x_{j_k} \in \mathcal{I}^{x_{j_k}}} \varphi_{i,k}(x_{j_k})$, and

$$\forall i \in \{1, \ldots n\}: \quad -\gamma_i \overline{x}_i + \overline{\lambda}_{i,1} \circ \ldots \circ \overline{\lambda}_{i,q_i} \le 0 \tag{4.6}$$

where $\overline{\lambda}_{i,k} = \max_{x_{j_k} \in \mathcal{I}^{x_{j_k}}} \varphi_{i,k}(x_{j_k})$.

Proof. Consider the hypersurface of the set $\mathcal{F}$ with $x_i = \underline{x}_i$. If Inequality (4.5) holds, the vector field is directed inward $\mathcal{F}$ everywhere on this hypersurface. Equivalently, the vector field is directed inward $\mathcal{F}$ on the hypersurface $x_i = \overline{x}_i$ if Inequality (4.6) holds. To show necessity, assume that at least one of these inequalities is not satisfied. Then, there exists a point x on a hypersurface of $\mathcal{F}$ such that the vector field is not directed inward or tangent to this hypersurface at that point x. $\square$

From this proposition we can conclude that the sets $\mathcal{F}_z$, $z = 1, \ldots, m$, are forward-invariant if and only if all monotonic functions $\varphi_{i,k}$ satisfy Proposition 4.7 for every set $\mathcal{F}_z$. The existence of a forward-invariant set $\mathcal{X}$ is intrinsic to the structure of the right-hand side functions f_i, as stated next.

Proposition 4.8. *Let $\bar{r}_i < \infty$ be the maximum value that the production term r_i can assume. Then, the set $\mathcal{X} = [0, x_1^{\max}] \times \ldots \times [0, x_n^{\max}]$, with $x_i^{\max} = \frac{\bar{r}_i}{\gamma_i}$, is forward-invariant.*

Proof. Assume $x_0 \in \mathcal{X}$. The trajectory $x(t; x_0)$ cannot cross any of the bounding hyperplanes $x_i = 0$ or $x_i = x_i^{\max}$ of $\mathcal{X}$, as $\dot{x}_i \ge 0$ if $x_i = 0$, and $\dot{x}_i \le 0$ if $x_i = x_i^{\max}$. $\square$

4.3. A combinatorial approach for the validation of multistability in gene regulation networks

In this section, we assume that several steady states of a gene regulation network have been observed experimentally, and that it is possible to classify the concentration of each species at each steady state as either "low" or "high", yet without knowing concentration values. In the terminology introduced in the last section, this means that only the ordering of the critical concentrations is known, that is, these variables satisfy $0 < x_i^{\text{low}} < x_i^{\text{high}} < x_i^{\text{max}}$. The hyperrectangular regions $\mathcal{X}$ and $\mathcal{F}_z$ are then defined as before. Furthermore, it is assumed that a hypothesis about the interaction structure of the gene regulation network is formulated as an uncertain model according to Equations (4.2) and (4.2), in which the exact shapes of the involved monotonic functions are not known.

The model validation problem is then to test if the hypothetical model structure can in principle generate the observed forward-invariant sets. Mathematically, it has to be verified that there exist degradation rates γ_i, critical concentrations x_i^{low}, x_i^{high}, and x_i^{max}, and monotonic functions $\varphi_{i,k}(x_j^{\text{low}})$ such that the sets $\mathcal{X}$ and $\mathcal{F}_z$, $z = 1, \ldots, m$, are forward-invariant.

In the following, we will reduce this validation problem to a combinatorial problem, which requires the fulfillment of a set of Boolean equations that can be generated from a number of atomic expressions. The result will be an efficient and easily applicable algorithm. Some definitions are given first.

4.3.1. Definitions

In case that no numeric values are specified for the variables x_i^{low}, x_i^{high} and x_i^{max}, we call the intervals $\mathcal{I}^{x_i,l}$ and $\mathcal{I}^{x_i,h}$ *qualitative* intervals. For these intervals, we define a Boolean value $B(\mathcal{I}^{x_i})$ according to

$$B(\mathcal{I}_x) = \begin{cases} 0 & \text{if } 0 \in \mathcal{I}^{x_i} \text{ and } x_i^{\text{max}} \notin \mathcal{I}^{x_i}, \\ 1 & \text{if } 0 \notin \mathcal{I}^{x_i} \text{ and } x_i^{\text{max}} \in \mathcal{I}^{x_i}, \\ \text{undefined} & \text{otherwise.} \end{cases} \tag{4.7}$$

This means that only low and high intervals have well-defined Boolean values. Based on this concept, generalized qualitative intervals and generalized critical values are defined.

Definition 4.9. *A generalized qualitative interval $\mathcal{I}^x := \mathcal{I}^{x_1} \times \ldots \times \mathcal{I}^{x_n}$ is the Cartesian product of intervals $\mathcal{I}^{x_i}$, $i = 1, \ldots, n$, all of which have well-defined Boolean values according to 4.7.*

In particular, Definition 4.9 states that all hyperrectangular sets $\mathcal{F}_z$, $z = 1, \ldots, m$, representing the observed steady states are generalized qualitative intervals. Next, the notion of admissible intervals is introduced.

Definition 4.10. *Consider an equation $\dot{x}_i = -\gamma_i x_i + r_i(x)$ from System (4.1). Furthermore, assume that a qualitative generalized interval $\mathcal{I}^x$ is given. Then, $\mathcal{I}^x$ is said to be admissible for this equation if the following holds: there exist degradation rates γ_i and critical concentrations*

x_i^{low}, x_i^{high}, x_i^{max}, $i = 1, \ldots, n$, as well as monotonic functions $\varphi_{i,1}, \ldots, \varphi_{i,q_i}$ such that all inequalities resulting from

$$\forall x \in \mathcal{I}^x : \dot{x}_i \geq 0 \text{ if } x_i = \min_{z \in \mathcal{I}^{x_i}} z \text{ and } \dot{x}_i \leq 0 \text{ if } x_i = \max_{z \in \mathcal{I}^{x_i}} z \tag{4.8}$$

are satisfied.

The following result is an immediate consequence of this definition.

Lemma 4.11. *Let $\mathcal{I}^x$ be a generalized interval. If $\mathcal{I}^x$ is admissible for all equations $\dot{x}_i = -\gamma_i x_i + r_i(x)$, $i = 1, \ldots, n$, then it is possible to choose all degradation rates γ_i, all critical concentrations x_i^{low}, x_i^{high}, x_i^{max}, $i = 1, \ldots, n$, and all monotonic functions $\varphi_{i,k}$ in System (4.1) such that $\mathcal{I}^x$ is forward-invariant.*

With Lemma 4.11, the model validation problem can now be reduced to the problem of verifying whether all sets $\mathcal{F}_z$, $z = 1, \ldots, m$, are admissible for all equations $\dot{x}_i = -\gamma_i x_i + r_i(x))$, $i = 1, \ldots, n$. We will next derive a systematic procedure to answer this question. Thereby, in order to verify if there exists monotonic activation or inhibition functions $\varphi_{i,k}$, we only check if there exist critical values for these functions that have the desired ordering. A function $\varphi_{i,k}$ can then for example be realized as piecewise linear function that assumes the critical values at the critical concentrations.

4.3.2. Combinatorial decision rules

The decision procedure consists of a finite set of decision rules which need to be satisfied. Let us first define mathematically, what we mean with a decision rule. The construction of decision rules for a given network is then presented in the remainder of this section.

Definition 4.12. *Given a differential equation*

$$\dot{x}_i = -\gamma_i x_i + \varphi_{i,1}(x_{j_1}) \circ \ldots \circ \varphi_{i,q_i}(x_{j_{q_i}}) \tag{4.9}$$

as defined in Equations (4.1) and (4.2), consider a Boolean equation of the form

$$B(\mathcal{I}^{x_i}) = \bar{B}(\mathcal{I}^x), \tag{4.10}$$

in which $\bar{B}(\mathcal{I}^x)$ is a Boolean expression assigning well-defined generalized intervals $\mathcal{I}^x$ the Boolean values 0 or 1. We call the Boolean Equation (4.10) a decision rule for Equation (4.9) if the following holds: whenever a generalized qualitative interval $\mathcal{I}^x = \mathcal{I}^{x_1} \times \ldots \times \mathcal{I}^{x_n}$ satisfies (4.10), then $\mathcal{I}^x$ is admissible for (4.9).

In other words, a decision rule gives a guarantee. If a generalized qualitative interval $\mathcal{I}^x$ satisfies a decision rule for each equation of the system of ordinary differential equations, then it is possible to choose the numeric values for the boundaries of $\mathcal{I}^x$ as well as the degradation rates and monotonic functions in the system such that $\mathcal{I}^x$ is forward-invariant.

Example 4.13. *The concept of decision rules is illustrated with the help of the mutual inhibition network from Example 4.5. As will be shown later in this section, one possible decision rule for the first equation of System (4.3) is*

$$B(\mathcal{I}^{x_1}) = \text{not } B(\mathcal{I}^{x_2}),$$

while one possible decision rule for the second equation is given by

$$B(\mathcal{I}^{x_2}) = \text{not } B(\mathcal{I}^{x_1}).$$

Consider now the generalized qualitative interval $\mathcal{I}^{x_1,l} \times \mathcal{I}^{x_2,h}$ and note that it satisfies both of the above decision rules. The same holds true for the generalized qualitative interval $\mathcal{I}^{x_1,h} \times \mathcal{I}^{x_2,l}$. This allows to conclude that there exist degradation rates γ_i and critical concentrations $x_i^{\text{low}} < x_i^{\text{high}} < x_i^{\text{max}}$, $i = 1, 2$, as well as inhibition functions $v_{1,1}(x_2)$ and $v_{2,1}(x_2)$ such that these two generalized qualitative intervals are indeed forward-invariant, which is in agreement with the phase portrait in Figure 4.2.

Decision rules for simple network structures

Let us start with the construction of decision rules for simple Equations (4.1), in which the production term is only a monotonic function $r_i(x) = \varphi(x_j)$. For this case, all possible decision rules are given in Table 4.1.

Table 4.1.: Elementary decision rules for simple production terms $r_i(x) = \varphi(x_j)$.

1.	$r_i(x) = v(x_j)$
i)	$B(\mathcal{I}^{x_i}) = B(\mathcal{I}^{x_j})$
ii)	$B(\mathcal{I})^{x_i} = 0$
2.	$r_i(x) = \mu(x_j)$
i)	$B(\mathcal{I}^{x_i}) = \text{not } B(\mathcal{I}^{x_j})$
ii)	$B(\mathcal{I}^{x_i}) = 1$

Lemma 4.14. *Table 4.1 contains all possible decision rules for Equation (4.1) if $r_i(x)$ consists of only one monotonic function, $r_i(x) = \varphi_i(x_j)$.*

Proof. To prove that Table 4.1 contains all possible decision rules for simple production terms, all possible Boolean functions $B(\mathcal{I}^{x_i}) = \bar{B}(\mathcal{I}^{x}j)$, which are only 4 in this case, have to be tested. We only show this for $r_i(x)$ being an activation function as the proof for $r_i(x)$ being an inhibition function follows the same arguments.

The four alternatives for $\bar{B}(\mathcal{I}^{x_j})$ are (i) $\bar{B}(\mathcal{I}^{x_j}) = 0$, (ii) $\bar{B}(\mathcal{I}^{x_j}) = B(\mathcal{I}^{x_j})$, (iii) $\bar{B}(\mathcal{I}^{x_j}) = \text{not } B(\mathcal{I}^{x_j})$, and (iv) $\bar{B}(\mathcal{I}^{x_j}) = 1$. Let us first consider (ii). In order for $B(\mathcal{I}^{x_i}) = B(\mathcal{I}^{x_j})$ to be a decision rule, the intervals $\mathcal{I}^{x_i,l} \times \mathcal{I}^{x_j,h}$ and $\mathcal{I}^{x_i,h} \times \mathcal{I}^{x_j,h}$ need to be admissible. From the condition in Equation (4.8) it then follows that the inequalities

$$v(0) \geq 0$$
$$-\gamma_i x_i^{\text{low}} + v(x_j^{\text{low}}) \leq 0$$
$$-\gamma_i x_i^{\text{high}} + v(x_j^{\text{high}}) \geq 0$$
$$-\gamma_i x_i^{\text{max}} + v(x_j^{\text{max}}) \leq 0$$

have to be satisfied. These inequalities in turn hold if $0 < \lambda^{\text{low}} = \nu(x_j^{\text{low}}) = \gamma_i x_i^{\text{low}} < \lambda^{\text{high}} = \nu(x_i^{\text{high}}) = \gamma_i x_i^{\text{high}} < \nu(x_j^{\text{max}}) \le M = \gamma_i x_i^{\text{max}}$ is true, which can easily be achieved. Next, note that the inequalities resulting from (i) can be satisfied by setting $M = \gamma_i x_i^{\text{low}}$. (iv) is no decision rule as this would require to satisfy the inequality $-\gamma_i x_i^{\text{high}} + \nu(0) \ge 0$, which however contradicts the definition of an activation function. Also for (iii), the resulting system of inequalities has no solution which respects the monotonicity properties of $\nu(x_j)$. $\qquad\qquad\qquad\qquad\qquad\square$

We can now proceed with the case that the production term consists of more than one monotonic function. Table 4.2 lists all possible decision rules for all multiplicative and additive combinations of two monotonic functions.

Lemma 4.15. *Table 4.2 contains all possible decision rules for Equation (4.1) if $r_i(x)$ consists of the additive or multiplicative composition of two monotonic functions, $r_i(x) = \varphi_{i,1}(x_{j_1}) \circ \varphi_{i,2}(x_{j_2})$.*

The proof of Lemma 4.16 is given in Appendix B.1.

Decision rules for general reaction structures

So far, we have derived all decision rules for production terms r_i which consist of either one or two monotonic functions. Based on these rules, we will now suggest a procedure to construct decision rules for more complex production terms $r_i(x)$.

Let now $\varphi_1(x)$ be a generalized activation or inhibition function and assume that $B(\mathcal{I}^{x_i}) = B_1(\mathcal{I}^x)$ is a decision rule for the equation $\dot{x}_i = -\gamma_i x_i + \varphi_1(x)$. In this, B_1 is a Boolean expression which assigns well-defined generalized intervals $\mathcal{I}^x$ the Boolean values 0 or 1. Furthermore, let φ_2 be a single activation or inhibition function and consider the differential equation $\dot{x}_i = -\gamma_i x_i + \varphi_1(x) \circ \varphi_2(x_k)$. The following lemma then states how B_1 and the elementary rules from Table 4.2 can be reused to obtain new decision rules for the concatenation $\varphi_1 \circ \varphi_2$. In the following, the symbol "$\diamond$" stands for either " " (nothing) or the logic "not", and the symbol "$\dagger$" stands for either the logic "and" or the logic "or".

Lemma 4.16. *Treat the generalized activation or inhibition function φ_1 as an activation function if $\varphi_1(0) = 0$, and as an inhibition function otherwise. In the case that φ_1 is interpreted as an activation function, let $B(\mathcal{I}^{x_i}) = \diamond B(\mathcal{I}^{x_j}) \dagger B(\mathcal{I}^{x_k})$ be a decision rule from Table 4.2 corresponding to $r_i(x) = \nu_1(x_j) \circ \varphi_2(x_k)$. Then, $B(\mathcal{I}^{x_i}) = B_1(\mathcal{I}^x) \dagger \diamond B(\mathcal{I}^{x_k})$ is a decision rule for $r_i(x) = \varphi_1(x) \circ \varphi_2(x_k)$. Analogously, in the case that φ_1 is interpreted as an inhibition function, let $B(\mathcal{I}^{x_i}) = \diamond B(\mathcal{I}^{x_j}) \dagger B(\mathcal{I}^{x_k})$ be a decision rule from Table 4.2 corresponding to $r_i(x) = \mu_1(x_j) \circ \varphi_2(x_k)$. Then, $B(\mathcal{I}^{x_i}) = B_1(\mathcal{I}^x) \dagger \diamond B(\mathcal{I}^{x_k})$ is a decision rule for $r_i(x) = \varphi_1(x) \circ \varphi_2(x_k)$.*

In words, this lemma states that we can substitute parts of a decision rule for easier structures with more complicated expressions. The proof of Lemma 4.15 is given in Appendix B.2. Note that the requirement that $\varphi_2(x_k)$ is an individual activation or inhibition function is important. Lemma 4.15 cannot be generalized to the case where φ_2 is a generalized activation or inhibition, that is, a composition of several activation or inhibition functions, too. An example showing why this is not possible is also given in Appendix B.3.

Table 4.2.: Elementary decision rules for composed production terms $r_i(x) = \varphi_{i,1}(x_{j_1}) + \varphi_{i,2}(x_{j_2})$ or $r_i(x) = \varphi_{i,1}(x_{j_1}) \cdot \varphi_{i,2}(x_{j_1})$.

3.	$r_i(x) = \nu_1(x_j) \cdot \nu_2(x_k)$
i)	$B(\mathcal{I}^{x_i}) = B(\mathcal{I}^{x_j})$ and $B(\mathcal{I}^{x_k})$
ii)	$B(\mathcal{I}^{x_i}) = 0$
4.	$r_i(x) = \nu_1(x_j) \cdot \mu_2(x_k)$
i)	$B(\mathcal{I}^{x_i}) = B(\mathcal{I}^{x_j})$ and $(\text{not } B(\mathcal{I}^{x_k}))$
ii)	$B(\mathcal{I}^{x_i}) = B(\mathcal{I}^{x_j})$
iii)	$B(\mathcal{I}^{x_i}) = 0$
5.	$r_i(x) = \mu_1(x_j) \cdot \mu_2(x_k)$
i)	$B(\mathcal{I}^{x_i}) = (\text{not } B(\mathcal{I}^{x_j}))$ and $(\text{not } B(\mathcal{I}^{x_k}))$
ii)	$B(\mathcal{I}^{x_i}) = (\text{not } B(\mathcal{I}^{x_j}))$ or $(\text{not } B(\mathcal{I}^{x_k}))$
iii)	$B(\mathcal{I}^{x_i}) = \text{not } B(\mathcal{I}^{x_j})$
iv)	$B(\mathcal{I}^{x_i}) = \text{not } B(\mathcal{I}^{x_k})$
v)	$B(\mathcal{I}^{x_i}) = 1$
6.	$r_i(x) = \nu_1(x_j) + \nu_2(x_k)$
i)	$B(\mathcal{I}^{x_i}) = B(\mathcal{I}^{x_j})$ and $B(\mathcal{I}^{x_k})$
ii)	$B(\mathcal{I}^{x_i}) = B(\mathcal{I}^{x_j})$ or $B(\mathcal{I}^{x_k})$
iii)	$B(\mathcal{I}^{x_i}) = B(\mathcal{I}^{x_j})$
iv)	$B(\mathcal{I}^{x_i}) = B(\mathcal{I}^{x_k})$
v)	$B(\mathcal{I}^{x_i}) = 0$
7.	$r_i(x) = \nu_1(x_j) + \mu_2(x_k)$
i)	$B(\mathcal{I}^{x_i}) = B(\mathcal{I}^{x_j})$ and $(\text{not } B(\mathcal{I}^{x_k}))$
ii)	$B(\mathcal{I}^{x_i}) = B(\mathcal{I}^{x_j})$ or $(\text{not } B(\mathcal{I}^{x_k}))$
iii)	$B(\mathcal{I}^{x_i}) = B(\mathcal{I}^{x_j})$
iv)	$B(\mathcal{I}^{x_i}) = \text{not } B(\mathcal{I}^{x_k})$
v)	$B(\mathcal{I}^{x_i}) = 0$
vi)	$B(\mathcal{I}^{x_i}) = 1$
8.	$r_i(x) = \mu_1(x_j) + \mu_2(x_k)$
i)	$B(\mathcal{I}^{x_i}) = (\text{not } B(\mathcal{I}^{x_j}))$ and $(\text{not } B(\mathcal{I}^{x_k}))$
ii)	$B(\mathcal{I}^{x_i}) = (\text{not } B(\mathcal{I}^{x_j}))$ or $(\text{not } B(\mathcal{I}^{x_k}))$
iii)	$B(\mathcal{I}^{x_i}) = \text{not } B(\mathcal{I}^{x_j})$
iv)	$B(\mathcal{I}^{x_i}) = \text{not } B(\mathcal{I}^{x_k})$
v)	$B(\mathcal{I}^{x_i}) = 1$

With this preliminary result, we can now present a procedure to construct decision rules for more complicated terms. Because of the last remark, we restrict ourselves to the case where $r_i(x)$ contains only summands or only products, that is, we do not allow productions terms in which both, sums and products occur. Thus, we consider production terms of the form

$$r_i(x) = \varphi_1(x_{j_1}) \circ \ldots \circ \varphi_p(x_{j_p}), \qquad (4.11)$$

in which the symbol "$\circ$" means either always "+" or always "$\cdot$". Note that we have also omitted the index i for the activation and inhibition functions compared to Equation (4.15). The procedure to construct the decision rules consists of several steps.

1) If $p = 1$, collect the Boolean expression $\diamond B(\mathcal{I}^{x_{j_1}})$ from all decision rules $B(\mathcal{I}^{x_i}) = \diamond B(\mathcal{I}^{x_{j_1}})$ from Table 4.1 corresponding to $r_i(x) = \varphi_1(x_{j_1})$ in a set $\mathcal{B}$. Proceed with step 3.

2) If $p \geq 2$,

   2a) consider the sum or product of the first two activation functions, $\varphi_1(x_{j_1}) \circ \varphi_2(x_{j_2})$. For every decision rule $B(\mathcal{I}^{x_i}) = \diamond B(\mathcal{I}^{x_{j_1}}) \dagger \diamond B(\mathcal{I}^{x_{j_2}})$ from Table 4.2 corresponding to this product or sum, add the Boolean expression $\diamond B(\mathcal{I}^{x_{j_1}}) \dagger \diamond B(\mathcal{I}^{x_{j_2}})$ to a set $\mathcal{B}$. If $p = 2$, continue with step 3. Otherwise, set $k = 3$ and proceed with step 2b.

   2b) Consider the first k summands or products $\varphi_1(x_{j_1}) \circ \ldots \circ \varphi_k(x_{j_k})$. Treat $\hat{\varphi}(\hat{x}) := \varphi_1(x_{j_1}) \circ \ldots \circ \varphi_{k-1}(x_{j_{k-1}})$ as activation function if $\hat{\varphi}(0) = 0$, and as inhibition otherwise. For every decision rule $B(\mathcal{I}^{x_i}) = \diamond B(\mathcal{I}^{\hat{x}}) \dagger \diamond B(\mathcal{I}^{x_{j_k}})$ from Table 4.2, that corresponds to the sum or product $r_i(x) = \hat{\varphi}(\hat{x}) + \varphi_k(x_{j_k})$, do the following: For every Boolean expression $\bar{B}$ from $\mathcal{B}$, create a new Boolean expression $\bar{B} \dagger \diamond B(\mathcal{I}^{x_{j_k}})$, replacing $\diamond B(\mathcal{I}^{\hat{x}})$ with $\bar{B}$. Collect these new Boolean expressions in a set $\bar{\mathcal{B}}$. After all possible new Boolean expressions have been created, replace the set $\mathcal{B}$ with $\bar{\mathcal{B}}$. If $k = p$, continue with step 3. Otherwise, set k to $k + 1$ and repeat step 2b.

3) Create the decision rule $B(\mathcal{I}^{x_i}) = \bar{B}$ for every Boolean expression $\bar{B}$ in $\mathcal{B}$.

Example 4.17. *Let us explain this procedure with the help of an example. Consider the equation*

$$\dot{x}_i = -\gamma_i x_i + \nu_1(x_1) \cdot \nu_2(x_2) \cdot \mu_3(x_3).$$

One of the decision rules that can be constructed for this equation is given below. The underbraces show which decision rule from Table 4.2 has been applied.

$$B(\mathcal{I}^{x_i}) = \underbrace{(\underbrace{B(\mathcal{I}^{x_1}) \text{ and } B(\mathcal{I}^{x_2})}_{3.i)}) \text{ and } (\text{not } B(\mathcal{I}^{x_2}))}_{4.i)}$$

The following result can be stated about the Boolean expressions created by this procedure.

Theorem 4.18. *All Boolean equations generated by the above procedure are decision rules for a production term $r_i(x)$ according to (4.2), in which "$\circ$" is always either "$+$" or "$\cdot$".*

Proof. This result follows from application of Lemma 4.2 to $\tilde{\varphi}_2(x) := \varphi_1(x_{j_1}) \circ \varphi_2(x_{j_2})$, and then iterative application of Lemma 4.16 to $\tilde{\varphi}_k := \tilde{\varphi}_{j_{k-1}}(x) \circ \varphi_k(x_{j_k})$, for $k = 3, \ldots, p$. □

So far, we have developed decision rules that allow to enumerate possible constellations of admissible intervals for an equation $\dot{x}_i = -\gamma_i x_i + r_i(x)$. For the case that $r_i(x)$ is composed of at most two activation or inhibition functions, Tables 4.1 and 4.2

are complete in the sense that they contain all possible decision rules. For the case that the symbol "∘" is always either "+" or always "·", the decision rules obtained by the procedure in the last section allow to enumerate at least a large number of possible constellations. More decision rules can be obtained, if permutations of the individual terms in Equation (4.11) are considered as well. We hypothesize, that all possible decision rules can be obtained by considering all possible such permutations.

4.3.3. An algorithmic solution to the model validation problem

We now include these decision rules into an algorithmic procedure, which solves the model validation problem as introduced at the beginning of Section 4.3 at least partially. As input to the algorithm, m generalized qualitative intervals need to be specified, for which we wish to decide if they can, in principle, be forward-invariant under the dynamics of the model. The output will be a simple "yes" or "no" answer. Again, we restrict ourselves to the case that each production term r_i consists of only sums or only products.

Algorithm 4.1: Validation of multistability.

Input: qualitative generalized intervals $\mathcal{F}_z$, $z = 1, \ldots, m$
Output: yes or no

1 create an empty set S for $i = 1, \ldots, n$ **do**
2   **foreach** *decision rule R_i for $\dot{x}_i = -\gamma_i \cdot x_i + r_i(x)$* **do**
3   **if** *all $\mathcal{I}_z^x$, $z = 1, \ldots, m$ are a solution to R_i* **then**
4   add i to S continue with $i := i + 1$

5 **if** *S contains all values $i = 1, \ldots, n$* **then**
6   **return** *yes*

7 **else**
8   **return** *no*

In line 2 of this algorithm, also all permutations of the individual activation or inhibition functions in a production term r_i should be considered. From Proposition 4.18 the main result for the model validation problem follows.

Theorem 4.19. *Let the generalized qualitative intervals $\mathcal{F}_z$, $z = 1, \ldots, m$, be the input to Algorithm 4.1. If it returns "no", then, for the case that no production term $r_i(x)$ is composed of more than two monotonic functions, it can be concluded that there do not exist degradation rates γ_i, monotonic functions $\varphi_{i,k}$ and values $0 < x_i^{\text{low}} < x_i^{\text{high}} < x_i^{\text{max}}$, such that the sets $\mathcal{F}_z$ are forward-invariant. If it returns "yes", the existence of degradation rates γ_i, monotonic functions $\varphi_{i,k}$ and values $0 < x_i^{\text{low}} < x_i^{\text{high}} < x_i^{\text{max}}$, such that the sets $\mathcal{F}_z$ are forward-invariant, is guaranteed.*

Proof. For the proof it is sufficient to see that each function $\dot{x}_i = -\gamma_i x_i + r_i(x)$ can be considered individually. This is the case as the numeric values of the variables x_i^{low}, x_i^{high}, x_i^{max}, $i = 1, \ldots, n$, play no role for the applicability of a decision rule for this equation. Then, Proposition 4.18 proves this result. $\qquad\square$

Complexity

The complexity of Algorithm 4.1 is briefly evaluated. Let us assume that each species in the network has at most k regulators. As, in the worst case of rule 7 in Table 4.2, there are 6 decision rules for a pair of monotonic functions, a production term with a given order of k activation and inhibition functions can maximally produce 6^{k-1} different rules. Moreover, there are $k!$ permutations of these activation and inhibition functions such that, in summary, there can be at most $k! \, 6^{k-1}$ decision rules. As each of the n equations of System (4.1) can be examined separately, the complexity is of order $n \cdot k! \, 6^{k-1}$. Thus, the algorithm is exponential in the maximal number of regulators per species, but linear in the system size. If the bound on the maximal number of regulators per species is dropped, the complexity grows exponentially with the system size.

4.3.4. Summary of this method

With Algorithm 4.1, we have developed a fast and easily applicable method to test, whether a certain model structure is, at least in principle, suited to reproduce an experimentally observed multistable behavior. The approach can be seen as bridge between a quantitative ODE and a qualitative Boolean description Boolean. It only relies on qualitative information and requires no detailed kinetic information or numeric values for the concentrations. The method is especially helpful in very early stages of the modeling process. An example of how it can be applied to accept or refute a modeling hypothesis in the presence of qualitative knowledge and measurements only is presented in Section 4.5.1. A disadvantage of this method is however, that it does not allow to incorporate quantitative information if it is available. Furthermore, the simple "yes" or "no" result of Algorithm 4.1 might be insufficient if we want to compare several model hypotheses or want to know how well-suited a system is to generate a specific multistable behavior. Nothing can be said about the regions of attraction or the robustness of the multistable behavior. Therefore, we extend this method in the next section such that it also allows to address these quantitative aspects.

4.4. Steady state robustness analysis and model discrimination

As in the last section, it is assumed that several steady states of a gene regulation network have been observed experimentally. Again, we want to treat each steady state as a hyperrectangular forward-invariant region in the state space. But in contrast to the last section, we now expect numeric values for the boundaries of each region. Instead of the purely qualitative answer, the method developed in this section then allows to decide if the system can exhibit the now quantitatively specified forward-invariant regions. Moreover, we will define a new robustness measure which characterizes the plausibility of the model structure by quantifying how fine the individual activation and inhibition functions in the network need to be tuned in order to achieve this multistable behavior. This robustness measure is thus the key to compare and rank different model hypotheses. Finally, we will show that this measure can be computed efficiently using convex optimization theory.

The first problem, the incorporation of numeric values for all variables x_i^{low}, x_i^{high}, and x_i^{max}, $i = 1, \ldots, n$, has already been reduced to a feasibility problem in Proposition 4.7. As it will also be part of the problem to compute the robustness measure studied later in this chapter, it is now sufficient to mention that it results in a convex feasibility problem, if every symbol "$\circ$" is either always "$+$" or always "$\cdot$". This will be discussed in detail later.

4.4.1. Development of the robustness measures

As outlined in the introduction of this chapter, we aim to develop a robustness measure for the interaction structure, that is, not for a specific choice of parameters or realizations of the activation or inhibition functions. In order to be able to draw conclusions that hold for a class of activation and inhibition functions and not only for one choice of them, we next introduce the concept of tubes for these monotonic functions, which is also illustrated in Figure 4.3.

Definition 4.20. *The 3-tuple of pairs of positive real numbers $T_{\mathcal{N}} = ((x^{\text{low}}, \tau^{\text{low}}), (x^{\text{high}}, \tau^{\text{high}}), (x^{\text{max}}, \tau^{\text{max}}))$ such that $\tau^{\text{low}} \leq \tau^{\text{high}} \leq \tau^{\text{max}}$ and $x^{\text{low}} \leq x^{\text{high}} \leq x^{\text{max}}$ is called a tube for activation functions. An activation function $\nu \in \mathcal{N}$ is said to satisfy a tube $T_{\mathcal{N}}$, denoted as $\nu \vDash T_{\mathcal{N}}$, if it holds*

$$\forall x \leq x^{\text{low}} : \nu(x) \leq \tau^{\text{low}}$$

$$\forall x \geq x^{\text{high}} : \nu(x) \geq \tau^{\text{high}}$$

$$\forall x \leq x^{\text{max}} : \nu(x) \leq \tau^{\text{max}}.$$

Equivalently, the 3-tuple of pairs of positive real numbers $T_{\mathcal{M}} = ((x^{\text{low}}, \tau^{\text{high}}), (x^{\text{high}}, \tau^{\text{low}}), (x^{\text{max}}, \tau^{\text{min}}))$ such that $\tau^{\text{min}} \leq \tau^{\text{low}} \leq \tau^{\text{high}}$ and $x^{\text{low}} \leq x^{\text{high}} \leq x^{\text{max}}$ is a called tube for inhibition functions. An inhibition function $\mu \in \mathcal{M}$ is said to satisfy a tube $T_{\mathcal{M}}$, denoted as $\mu \vDash T_{\mathcal{M}}$, if it holds

$$\forall x \leq x^{\text{low}} : \mu(x) \geq \tau^{\text{high}}$$

$$\forall x \geq x^{\text{high}} : \mu(x) \leq \tau^{\text{low}}$$

$$\forall x \leq x^{\text{max}} : \mu(x) \geq \tau^{\text{min}}.$$

If these inequalities are not satisfied, we write $\nu \nvDash T_{\mathcal{N}}$, or $\mu \nvDash T_{\mathcal{M}}$, respectively. Furthermore, we use T as abbreviation for both, $T_{\mathcal{N}}$ and $T_{\mathcal{M}}$. Considering a system as in (2.3), we assign a tube to each monotonic function. We will index a tube for $\varphi_{i,k}$ as $T^{i,k}$. Proposition 4.7 can now easily be restated in terms of tubes. To this end, consider a region $\mathcal{F} = \mathcal{I}^{x_1} \times \ldots \times \mathcal{I}^{x_n}$, where each interval is either $[0, x_i^{\text{low}}]$ or $[x_i^{\text{high}}, x_i^{\text{max}}]$, and define the following values for a tube $T^{i,k}$ for a function $\varphi_{i,k}(x_j)$.

$$\underline{\tau}_{i,k} = \begin{cases} 0 & \text{if } 0 \in \mathcal{I}^{x_j} \wedge \varphi_{i,k} \in \mathcal{N} \\ \min\{\tau : (x, \tau) \in T^{i,k} \wedge x_j \in \mathcal{I}^{x_j}\} & \text{otherwise} \end{cases} \tag{4.12}$$

and

$$\overline{\tau}_{i,k} = \max\{\tau : (x, \tau) \in T^{i,k} \wedge x \in \mathcal{I}^{x_j}\}. \tag{4.13}$$

In words, this definition states that $\underline{\tau}_{i,k}$ is the largest tube parameter τ which is a lower bound for $\varphi_{i,k}(x_j)$, $x_j \in \mathcal{I}^{x_j}$. Equivalently, $\overline{\tau}_{i,k}$ is the smallest tube parameter τ which is an upper bound for $\varphi_{i,k}(x_j)$, $x_j \in \mathcal{I}^{x_j}$.

Proposition 4.21. *Given a set $\mathcal{F} = \mathcal{I}^{x_1} \times \ldots \times \mathcal{I}^{x_n}$ such that $\mathcal{I}^{x_i} = [0, x_i^{\mathrm{low}}]$ or $\mathcal{I}^{x_i} = [x_i^{\mathrm{high}}, x_i^{\mathrm{max}}]$. Furthermore, given tubes $T^{i,k}$ that satisfy the conditions*

$$\forall i \in \{1, \ldots n\} : \; -\gamma_i \cdot \underline{x}_i + \underline{\tau}_{i,1} \circ \ldots \circ \underline{\tau}_{i,q_i} \geq 0 \tag{4.14}$$

where $\underline{x}_i = \min_{z \in \mathcal{I}^{x_i}} z$, and $\underline{\tau}_{i,k}$ as defined in (4.12), and

$$\forall i \in \{1, \ldots n\} : \; -\gamma_i \cdot \overline{x}_i + \overline{\tau}_{i,1} \circ \ldots \circ \overline{\tau}_{i,q_i} \leq 0 \tag{4.15}$$

where $\overline{x}_i = \max_{z \in \mathcal{I}^{x_i}} z$, and $\overline{\tau}_{i,k}$ as defined in (4.13). If $\forall i, k : \; \varphi_{i,k} \vDash T^{i,k}$, then the set $\mathcal{F}$ is forward-invariant under the dynamics of System (4.1).

Proof. Note that $\varphi_{i,k} \vDash T^{i,k}$ means that $\underline{\tau}_{i,k}$ is a lower bound on $\underline{\lambda}_{i,k}$, and $\overline{\tau}_{i,k}$ is an upper bound on $\overline{\lambda}_{i,k}$, with $\underline{\lambda}_{i,k}$ and $\overline{\lambda}_{i,k}$ as defined in Proposition 4.7. Therefore, if the Inequalities (4.14) and (4.15) hold for a tube $T^{i,k}$ and a set $\mathcal{F}$, then Equations (4.5) and (4.6) hold for all monotonic functions $\varphi_{i,k} \vDash T^{i,k}$, and the set $\mathcal{F}$ is thus forward-invariant. $\qquad \square$

The other direction is however not necessarily true. That is, given tubes $T^{i,k}$ which satisfy inequalities (4.14) and (4.15), it might be possible to find activation and inhibition functions $\varphi_{i,k}$, such that at least for some of these functions it holds that $\varphi_{i,k} \nvDash T^{i,k}$, but $\mathcal{F}$ is still forward-invariant. Therefore, Proposition 4.21 is sufficient but not necessary for the forward-invariance of $\mathcal{F}$.

Quantification of perturbations and definition of the robustness measure

For the definition of the robustness measure, it is necessary to define and quantify the perturbations against which the system should be robust. Here, we want to consider variations of the activating and inhibiting regulatory interactions as this includes many relevant classes of perturbations. Among others, perturbations by mutations in the promoter regions, environmental influences such as temperature or chemicals, or inherent stochastic fluctuations can be described in this way. For the quantification of perturbations, the l_1-norm will be used. More precisely, given a monotonic function $\varphi(x_j)$ and a perturbed function $\varphi^p(x_j) \in S_\varphi$,

$$\|\varphi - \varphi^p\|_1 = \int_0^\infty |\varphi(x_j) - \varphi^p(x_j)| \, \mathrm{d}x_j \tag{4.16}$$

is proposed as a measure for the perturbation of φ. To ensure that this integral exists, we furthermore impose the technical assumption that φ^p approaches φ sufficiently, that is, exponentially fast as x_j goes to infinity. As a further preparing step, let

$$\mathcal{R}^{\mathrm{min}}(\varphi, T) = \inf_{\varphi^p \in S_\varphi \wedge \varphi^p \nvDash T} \|\varphi - \varphi^p\|_1 \tag{4.17}$$

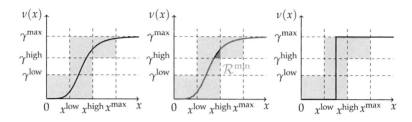

Figure 4.3.: Illustration of a tube for an activation function, the minimal perturbation $\mathcal{R}^{\min}$ as defined in (4.18), and the best centered step function which maximizes $\mathcal{R}^{\min}$.

be the minimal perturbation of a given function $\varphi \vDash T$ such that it violates T, $\varphi^p \nvDash T$ (see Figure 4.3 for an illustration). Also, define

$$\mathcal{R}^{\max}(T) = \sup_{\varphi \vDash T} \mathcal{R}^{\min}(\varphi, T). \tag{4.18}$$

If this supremum is achieved by a function $\tilde{\varphi}$, this function can be viewed as best centered with respect to the tube T. As will be shown in Section 4.4.2, $\mathcal{R}^{\max}(T)$ can be computed analytically. With this, the robustness measure for an interaction structure can be defined.

Definition 4.22. *Given a system according to Equation (4.1) with production terms r_i according to Equation (4.2). The robustness measure $\mathcal{R}$ for this system with respect to the forward-invariance of sets $\mathcal{F}_z$, $z = 1, \ldots, m$, and perturbations of the activation and inhibition functions measured by the l_1 norm, is given by*

$$\mathcal{R} = \max_{T^{i,k}} \min_{i,k} \mathcal{R}^{\max}(T^{i,k})$$

$$\text{s.t.: all tubes } T^{i,k} \text{ satisfy the conditions given in Proposition 4.21} \tag{4.19}$$

This measure $\mathcal{R}$ can be interpreted as a guarantee. It involves the computation of an optimal system, consisting of optimal tubes $\tilde{T}^{i,k}$ and monotonic functions $\tilde{\varphi}_{i,k}$ such that $\mathcal{R}$ is maximized. Then, for this optimal system it can be guaranteed that all sets $\mathcal{F}_z$ are still forward-invariant if no function $\tilde{\varphi}_{i,k}$ is perturbed by more than $\mathcal{R}$, that is, $\forall i, k : \|\tilde{\varphi}_{i,k} - \tilde{\varphi}_{i,k}^p\|_1 \leq \mathcal{R}$. This optimal system is designed such that the smallest value $\mathcal{R}^{\max}(T^{i,k})$ among all tubes is maximized. This is reasonable as the tube with the smallest value $\mathcal{R}^{\max}(T^{i,k})$ can be considered as the most fragile interconnection of the network and therefore determines the robustness of the whole network.

4.4.2. Computation of the robustness measure

We now develop a computationally attractive method to compute the robustness measure $\mathcal{R}$ as defined above. Therefore, let us first derive an analytic solution for $\mathcal{R}^{\max}$ as in Equation (4.18).

Proposition 4.23. *Let T_N be a tube for an activation function $\nu(x_j)$. Then, the maximal value $\mathcal{R}^{\max}(T_N)$ is given by*

$$\mathcal{R}^{\max}(T_N) = \frac{\tau^{\text{low}} \cdot (\tau^{\max} - \tau^{\text{high}})}{\tau^{\text{low}} + (\tau^{\max} - \tau^{\text{high}})} \cdot (x_j^{\text{high}} - x_j^{\text{low}}). \tag{4.20}$$

Equivalently, let T_M be a tube for an inhibition function $\mu(x_j)$. Then, $\mathcal{R}^{\max}(T_M)$ is given by

$$\mathcal{R}^{\max}(T_M) = \frac{(M - \tau^{\text{high}}) \cdot (\tau^{\text{low}} - \tau^{\min})}{(M - \tau^{\text{high}}) + (\tau^{\text{low}} - \tau^{\min})} \cdot (x_j^{\text{high}} - x_j^{\text{low}}). \tag{4.21}$$

This proposition is proved in Appendix B.4. As illustrated in Figure 4.3, $\mathcal{R}^{\max}$ is achieved a best centered step function. With this result and the new variables

$$c_{i,k}^h = \begin{cases} \tau_{i,k}^{\max} - \tau_{i,k}^{\text{high}} & \text{if } \varphi_{i,k} \text{ is an activation function} \\ M_{i,k} - \tau_{i,k}^{\text{high}} & \text{if } \varphi_{i,k} \text{ is an inhibition function} \end{cases}$$

$$c_{i,k}^l = \begin{cases} \tau_{i,k}^{\text{low}} & \text{if } \varphi_{i,k} \text{ is an activation function} \\ \tau_{i,k}^{\text{low}} - \tau_{i,k}^{\min} & \text{if } \varphi_{i,k} \text{ is an inhibition function,} \end{cases} \tag{4.22}$$

Problem (4.19) can be rewritten as

$$\mathcal{R} = \max_{T^{i,k}} \min_{i,k} \left\{ \frac{c_{i,k}^h \cdot c_{i,k}^l}{c_{i,k}^h + c_{i,k}^l} (x_{i,k}^{\text{high}} - x_{i,k}^{\text{low}}) \right\}$$

s.t. all tubes $T^{i,k}$ satisfy the conditions given in Proposition 4.21

all equalities from Equation (4.22) hold. $\tag{4.23}$

This in turn can be reformulated as

$$\mathcal{R} = \min_{T^{i,k}} \frac{1}{t}$$

s.t.: $\forall i, k: \ t \leq \left\{ \frac{c_{i,k}^h \cdot c_{i,k}^l}{c_{i,k}^h + c_{i,k}^l} (x_{i,k}^{\text{high}} - x_{i,k}^{\text{low}}) \right\}$ $\tag{4.24}$

all tubes $T^{i,k}$ satisfy the conditions given in Proposition 4.21

all equalities from Equation (4.22) hold.

The optimization problem of Equation (4.24) still involves minimization over all admissible tubes. This can be made explicit by adding constraints on the τ values of the individual tubes. These additional constraints are derived next.

First, the tubes have to satisfy the maximum concentration constraints $x_i^{\max}$ for each protein. From the desired forward-invariance of the set $\mathcal{X} = [0, x_i^{\max}] \times \ldots \times [0, x_n^{\max}]$ and application of Proposition 4.7 to $\mathcal{X}$, the conditions

$$- \gamma_i \cdot x_i^{\max} + \hat{\tau}_{i,1} \circ \ldots \circ \hat{\tau}_{i,q_i} \leq 0 \quad i = 1, \ldots, n, \tag{4.25}$$

are obtained, in which $\hat{\tau}_{i,k} = M_{i,k}$ if $\varphi_{i,k}$ is an inhibition function and $\hat{\tau}_{i,k} = \tau_{i,k}^{\max}$ if it is an activation function.

Secondly, the monotonicity constraints from the definition of the tubes have to be considered. Therefore, for every tube $T_{\mathcal{N}}^{i,k}$, the constraint

$$0 \leq \tau_{i,k}^{\text{low}} \leq \tau_{i,k}^{\text{high}} \leq \tau_{i,k}^{\text{max}} \tag{4.26}$$

and for every tube $T_{\mathcal{M}}^{i,k}$, the constraint

$$0 \leq \tau_{i,k}^{\text{min}} \leq \tau_{i,k}^{\text{low}} \leq \tau_{i,k}^{\text{high}} \leq M_{i,k} \tag{4.27}$$

has to be included. Finally, we require all optimization variables to be positive. Collecting all optimization variables in a vector X, the resulting optimization problem can then be written as

$$\mathcal{R} = \min_{X} \frac{1}{t}$$

$$\text{s.t.: } \forall i,k : \ t \leq \left\{ \frac{c_{i,k}^{h} \cdot c_{i,k}^{l}}{c_{i,k}^{h} + c_{i,k}^{l}} \left(x_{i,k}^{\text{high}} - x_{i,k}^{\text{low}} \right) \right\}$$

s.t.: all tubes $T^{i,k}$ satisfy the conditions given in Proposition 4.21,

all equalities from Equation (4.22) hold,

all inequalities from Equations (4.25), (4.26), and (4.27) hold,

$X \geq 0$. $\tag{4.28}$

Formulation as a convex optimization problem

In this section, we aim to formulate Problem 4.28 as a convex problem, whose definition is recalled next for completeness.

Definition 4.24. *A convex optimization problem has the standard form*

$$\begin{aligned} \min \quad & f_0(x) \\ \text{s.t.} \quad & f_i(x) \leq 0, \quad i = 1, \ldots, m \\ & h_i(x) = 0, \quad i = 1, \ldots, p \end{aligned} \tag{4.29}$$

where the objective function $f_0 : \mathbb{R}^n \to \mathbb{R}$, and the inequality constraints $f_i : \mathbb{R}^n \to \mathbb{R}$, $i = 1, \ldots, m$, are convex functions, and the equality constraints $h_i : \mathbb{R}^n \to \mathbb{R}$ are affine in x.

Unfortunately, it will not always be possible to transform Problem (4.28) into an equivalent convex formulation, but we will identify two cases where these optimization problems are already given or can easily be transformed into a convex form. These two cases are studied next.

Additive combinations of monotonic functions

In the first case, each symbol "$\circ$" in a production term r_i according to (4.2) represents an addition. The following result can be stated.

Proposition 4.25. *Assume that each production term r_i of System (4.1) is an additive combination of activation or inhibition functions. Then, the optimization problem from Equation (4.28) is convex.*

Proof. It has to be checked that all requirements from Definition 4.24 are fulfilled. Clearly, the objective $f_0 = \frac{1}{t}$ is convex for $t > 0$. Also, the constraints $t - (\frac{c_{i,k}^h \cdot c_{i,k}^l}{c_{i,k}^h + c_{i,k}^l}(x_{i,k}^{\text{high}} - x_{i,k}^{\text{low}})) \leq 0$ are convex. To see this, the second-order condition (Boyd & Vandenberghe, 2004) can be applied. The eigenvalues of the Hessian of this constraint are computed as $\{0, 0, \frac{2((c_{i,k}^h)^2 + (c_{i,k}^l)^2)}{((c_{i,k}^h)^2 + (c_{i,k}^l)^2)^3}\}$. Therefore, the Hessian is positive semi-definite on the domain $(c_{i,k}^h, c_{i,k}^l, t) \in \mathbb{R}_+^3$ and the constraint is thus convex on this domain. Furthermore, the equality constraints according to (4.22) as well as the inequality constraints according to (4.14), (4.15), (4.25), (4.26), and (4.27) are affine in the optimization variables and thus convex. Thus, the objective as well as all constraints satisfy the respective requirements for $X \geq 0$. $\square$

Multiplicative combinations of monotonic functions

In the second case, all symbols "$\circ$" stand for a multiplication. If the equality constraints from Equation (4.22) are changed into the inequality constraints

$$
\begin{aligned}
c_{i,k}^h &\leq \begin{cases} \tau_{i,k}^{\max} - \tau_{ij}^{\text{high}} & \text{if } \varphi_{i,k} \text{ is an activation function} \\ M_{i,k} - \tau_{i,k}^{\text{high}} & \text{if } \varphi_{i,k} \text{ is an inhibition function} \end{cases} \\
c_{i,k}^l &\leq \begin{cases} \tau_{i,k}^{\text{low}} & \text{if } \varphi_{i,k} \text{ is an activation function} \\ \tau_{i,k}^{\text{low}} - \tau_{i,k}^{\min} & \text{if } \varphi_{i,k} \text{ is an inhibition function,} \end{cases}
\end{aligned}
\tag{4.30}
$$

the following result can be given.

Proposition 4.26. *Assume that each production term r_i of System (4.1) is a multiplicative combination of activation or inhibition functions. Then, the optimization problem from Equation (4.28), with the relaxed constraints from Equation (4.30) instead of Equation (4.22), is convex. Moreover, this relaxation does not change the optimal value.*

In order to prove this proposition, the definitions of monomials, posynomials and of a geometric program are needed, which we recall next from Boyd & Vandenberghe (2004) for convenience.

Definition 4.27. *A function $f : \mathbb{R}^n \to \mathbb{R}$ with domain $\mathbb{R}_+^n$ of the form $f(x) = cx_1^{a_1}x_2^{a_2}\ldots x_n^{a_n}$, $c > 0$, and $a_i \in \mathbb{R}$ is a monomial. A finite sum of monomials $F(x) = \sum_{k=1}^K f_k(x)$, $K \in \mathbb{N}_+$, is called a posynomial.*

Definition 4.28. *An optimization problem*

$$
\begin{aligned}
\min \quad & f_0(x) \\
\text{s.t.} \quad & f_i(x) \leq 1, \quad i = 1,\ldots,m \\
& h_i(x) = 1, \quad i = 1,\ldots,p
\end{aligned}
\tag{4.31}
$$

with domain $\mathbb{R}_+^n$, where f_0 and f_i, $i = 0,\ldots,m$, are posynomials, and h_i, $i = 1,\ldots,p$, are monomials is called a geometric program.

A geometric program can be transformed into a convex problem of the form of Equation (4.29) by the variable transformation $y_i = \log x_i$ (Boyd & Vandenberghe, 2004). With this, Proposition 4.26 can now be proved.

Proof. Convexity: We show that the optimization problem is given as a geometric program. First, the objective $f_0 = \frac{1}{t}$ is a posynomial. Also, the constraints $t - \frac{c_{i,k}^h \cdot c_{i,k}^l}{c_{i,k}^h + c_{i,k}^l}(x_{i,k}^{\text{high}} - x_{i,k}^{\text{low}}) \leq 0$ can be reformulated as posynomial $t \cdot a^{-1} \cdot (c_{i,k}^l)^{-1} + t \cdot a^{-1} \cdot (c_{i,k}^h)^{-1} \leq 1$, with constant $a = (x_{i,k}^{\text{high}} - x_{i,k}^{\text{low}})$. In an equivalent way, the inequality constraints from (4.14), (4.15), (4.25), (4.26), and (4.27) can be rewritten as posynomials. This is furthermore possible for the relaxed constraints from (4.30). Therefore, the optimization problem is given as a geometric problem.

Equivalence: First note that the feasible set of the original problem is contained in the feasible set of the relaxed problem. Denote $p_{\text{rel}}^\star$ the optimal value of the relaxed problem. Then it holds that $p_{\text{rel}}^\star \geq p^\star$. It now has to be shown that the optimal value for the relaxed problem is obtained when all relaxed inequalities (4.30) are satisfied with equality, that is, $p_{\text{rel}}^\star = p^\star$. Denote X the vector containing all optimization variables. It takes the value $X^\star$ at the optimal point $p^\star$. Recall that the optimal value $p^\star$ equals the smallest value $\mathcal{R}_{i,k}^{\max}$ of all tubes $T^{i,k}$ in the system. Then, all tubes $T^{i,k}$ in the system can be partitioned into two sets. The first set V contains all tubes for which it holds that $\mathcal{R}_{i,k}^{\max} = p^\star$ and the second set W contains all remaining tubes. For these tubes we have that $\mathcal{R}_{i,k}^{\max} > p_{\text{rel}}^\star$. Now assume that at $X = X^\star$ there is a relaxed constraint (4.30) which does not hold with equality. Then, without influencing any other constraint, equality of this constraint can be achieved by increasing the respective value $c_{i,k}^h$ or $c_{i,k}^l$. Then, as $\mathcal{R}_{i,k}^{\max}$ is monotonically increasing in $c_{i,k}^h$ and $c_{i,k}^l$, this value will also increase. As $p_{\text{rel}}^\star$ is optimal, there has to be at least one tube $T^{i,k}$ in the set V for which it holds that all constraints (4.30) hold with equality. Then modifying all other constraints which do not hold with equality in the way described above yields an optimal solution $p^\star$ for the original problem with $p^\star = p_{\text{rel}}^\star$. $\qquad\square$

4.4.3. Summary of this method

With the robustness measure $\mathcal{R}$ according to (4.19), we have defined a measure for the robustness of an interaction structure with respect to a desired multistable behavior. Unlike common ODE-based approaches, we do not require a fully parametrized system and the emphasis is thus more on the structure itself than on a specific set of parameters. This measure can thus be used to compare different hypotheses on the interaction structure, while other common ODE methods only allow for a comparison of different parameter sets for the same model. In contrast to the purely qualitative approach in Section 4.3, numeric values for the boundaries of the hyperrectangular forward-invariant sets are required now. If available, also kinetic information can be incorporated by setting for example tube constraints to fixed values. In addition to the definition of this robustness measure we have shown that it can be computed efficiently by solving a convex optimization problem if only additive or only multiplicative combinations of monotonic functions are considered. While requiring more quantitative information as the purely qualitative method developed in Section 4.3, the method is still intended to support modelers at early states of the modeling process. It allows to compare and evaluate various model structures that are all, at least in principle, capable of explaining the observed steady state pattern. Its use and benefits are demonstrated in the following section.

4.5. Application examples

We apply the two methods developed in this chapter to two example systems. In Section 4.5.1, we demonstrate with the help of a model of the lactose utilization network that the method for validation of multistability from Section 4.3 can be used to accept or refute hypothetical model structures based on qualitative information only, which is especially helpful in very early states of the modeling process. In Section 4.5.2, we apply the method for steady state robustness analysis and model discrimination from Section 4.4 to several minimal models for a binary decision step during stem cell differentiation. This example will show that the definition of the robustness measure is indeed suitable for biological networks. We will reveal several structural characteristics of such a network which render it maximally robust and thus, according to our interpretation, biologically plausible.

4.5.1. Multistability in the lactose utilization network

The lactose utilization network of Escherichia coli is essential for the bacterium to decide whether glucose or lactose is metabolized, whereby glucose is generally preferred. A mathematical model for this network has been proposed and analyzed by Ozbudak *et al.* (2004), and its most important species and their interplay are well understood: the lac operon codes for three genes, LacA, LacZ and LacY, which are responsible for uptake and metabolism of lactose and related sugars. Here, only LacY, which codes for lactose permease is considered, which regulates the uptake of lactose and similar molecules into the cell. One of these molecules is thio-methylgalactoside (TMG) which cannot be metabolized. In the model, an extracellular (eTMG) and an intracellular (iTMG) concentration are distinguished. A further protein, which is considered in the model, is cyclic AMP receptor protein (CRP), which is an activator for the lac operon. If glucose is present, the expression of CRP is inhibited. Furthermore, the model also includes a repressor of the lac operon, LacI, which can be inhibited by iTMG. Finally, glucose uptake inhibits the uptake of TMG. These findings are summarized in Figure 4.4 and can also be formulated with the mathematical formalisms presented in Section 4.2. A qualitative ODE model of this gene regulation network is given by

$$\dot{\text{LacY}} = -\gamma_1 \cdot \text{LacY} + \nu_{1,1}(\text{CRP}) + \mu_{1,2}(\text{LacI}) \tag{4.32}$$

$$\dot{\text{LacI}} = -\gamma_2 \cdot \text{LacI} + \nu_{2,1}(\text{iTMG}) \tag{4.33}$$

$$\dot{\text{iTMG}} = -\gamma_3 \cdot \text{iTMG} + \nu_{3,1}(\text{eTMG}) \cdot \nu_{3,2}(\text{LacY}) \cdot \mu_{3,3}(\text{Glu}) \tag{4.34}$$

$$\dot{\text{CRP}} = -\gamma_4 \cdot \text{CRP} + \mu_{4,1}(\text{Glu}), \tag{4.35}$$

where all $\nu_{i,k}$ are activation functions, and all $\mu_{i,k}$ are inhibition functions. In Equation (4.32), the combined influence of CRP and LacI on LacY is modeled by a sum of two monotonic functions as they can regulate the expression of LacY independently. Glu and eTMG are external inputs. The following experimental observations were reported by Ozbudak *et al.* (2004). In absence of glucose and for a low concentration of eTMG, the amount of LacY remains low, while for high concentrations of eTMG, LacY will be highly expressed. For intermediate eTMG concentrations, the system shows a hysteretic behavior. If glucose is present, the system still shows the same qualitative behavior, but the bifurcation points are shifted toward higher eTMG concentrations.

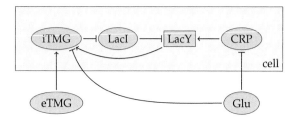

Figure 4.4.: Interaction structure of the lactose utilization network in Escherichia coli.

These findings are graphically summarized in the bifurcation diagram in Figure 4.5. Following the experimental observations, we require our model to exhibit the four forward-invariant regions

$$\mathcal{F}_1 = \mathcal{I}^{Glu,l} \times \mathcal{I}^{CRP,h} \times \mathcal{I}^{eTMG,l} \times \mathcal{I}^{iTMG,l} \times \mathcal{I}^{LacI,h} \times \mathcal{I}^{LacY,l}$$
$$\mathcal{F}_2 = \mathcal{I}^{Glu,l} \times \mathcal{I}^{CRP,h} \times \mathcal{I}^{eTMG,h} \times \mathcal{I}^{iTMG,h} \times \mathcal{I}^{LacI,l} \times \mathcal{I}^{LacY,h}$$
$$\mathcal{F}_3 = \mathcal{I}^{Glu,h} \times \mathcal{I}^{CRP,l} \times \mathcal{I}^{eTMG,l} \times \mathcal{I}^{iTMG,l} \times \mathcal{I}^{LacI,h} \times \mathcal{I}^{LacY,l}$$
$$\mathcal{F}_4 = \mathcal{I}^{Glu,h} \times \mathcal{I}^{CRP,l} \times \mathcal{I}^{eTMG,h} \times \mathcal{I}^{iTMG,l} \times \mathcal{I}^{LacI,h} \times \mathcal{I}^{LacY,l}.$$

An analysis with the combinatorial method from Section 4.3 shows that the system can indeed reproduce this qualitative multistable behavior. A proof for this conclusion are the following four Boolean equations, which result from the procedure described

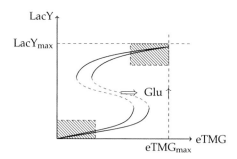

Figure 4.5.: Qualitative bifurcation diagram for the lactose utilization network. The figure shows the location of LacY at steady state for a fixed input eTMG. For larger values of the second input, Glu, the steady state curve is shifted to the right while staying qualitatively the same.

in Section 4.18. All four sets $\mathcal{F}_z$, $z = 1, \ldots, m$, are solutions to these equations.

$$
\begin{aligned}
B(\mathcal{I}^{\text{LacY}}) &= \text{not } B(\mathcal{I}^{\text{LacI}}) \\
B(\mathcal{I}^{\text{CRP}}) &= \text{not } B(\mathcal{I}^{\text{Glu}}) \\
B(\mathcal{I}^{\text{LacI}}) &= \text{not } B(\mathcal{I}^{\text{iTMG}}) \\
B(\mathcal{I}^{\text{iTMG}}) &= B(\mathcal{I}^{\text{eTMG}}) \text{ and } B(\mathcal{I}^{\text{LacY}}).
\end{aligned}
$$

Let us also give a brief example of how this method can be used for the invalidation of a model structure. Assume that it had been observed that intracellular TMG remains low in presence of lactose, even if extracellular TMG is high, that is, $\mathcal{F}_4$ is replaced by $\mathcal{F}_4' = \mathcal{I}^{\text{Glu,h}} \times \mathcal{I}^{\text{CRP,l}} \times \mathcal{I}^{\text{eTMG,h}} \times \mathcal{I}^{\text{iTMG,l}} \times \mathcal{I}^{\text{LacI,l}} \times \mathcal{I}^{\text{LacY,h}}$. Then, no decision rules can be found such that all all four sets, $\mathcal{F}_z$, $z = 1, \ldots, 3$, and $\mathcal{F}_4'$ are solutions to these Boolean equations. In particular, there is no such decision rule for Equation (4.33). As the production of this equation consists of only one monotonic function, the non-existence of decision rules for that case would be sufficient to refute the qualitative ODE model.

In summary, this example shows that the method for qualitative validation of multistability is an easy and fast test to accept or refuse a hypothetical interaction structure of a gene regulation network in the presence of qualitative knowledge and qualitative measurements only. In this fashion, hypothetical model structures that are invalidated with this method can quickly be sorted out and a more in-depth analysis can follow for those model structures only, that have passed this first test.

4.5.2. Three-node-networks for cell differentiation

Stem cells and their potential to give rise to multiple cell types have become a focus in systems biology and mathematical modeling throughout the last years (Ben D. MacArthur & Lemischka, 2009; Peltier & Schaffer, 2010). Starting from a multipotent state, stem cells undergo the process of cell differentiation which is completed when the cell ends up in a mature cell state. In the literature, cell differentiation is often viewed as a sequence of mostly binary cell fate decisions (Foster *et al.*, 2009). Figure 4.6 illustrates this with the example of osteoblasts. Thereby, one generic module consisting of only few genes is considered responsible for each of these decision steps (Xiong & Ferrell Jr., 2003; S. Huang & Enver, 2007; Foster *et al.*, 2009; Schittler *et al.*, 2010).

From studies of specific stem cells, such as hematopoietic or mesenchymal stem cells, one can identify common properties and motifs of these modules. For example, and as already emphasized in Section 4.1, distinct cell types are represented by distinct stable steady states. Each of the binary decision module thus needs to exhibit at least three steady states. However, the exact network structure network of the decision modules are mostly unknown.

In this example, we therefore perform an elementary study of small network motifs in the spirit of Klemm & Bornholdt (2005), Prill *et al.* (2005), and Rodrigo & Elena (2011). These motifs consist of only three genes, one associated to each of the three cell states, that is, progenitor, first and second differentiated cell type, which are involved in a decision step. As this is a very basic study, we use the qualitative modeling framework introduced in Section 4.2. Interpreting each steady state as hyperrectangular region in the state space of this network, we apply the method for model steady state robustness

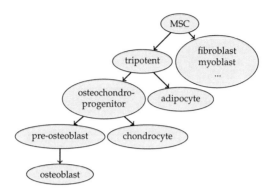

Figure 4.6.: Sequence of binary decision steps for the differentiation of mesenchymal stem cells (MSC) into osteoblasts.

analysis and model discrimination developed in Section 4.4 to uncover structural properties of these three-node networks, which allow the maintenance of these steady states in a maximally robust way, and might thus be favored in real networks, too.

Requirements on a differentiation motif

As mentioned above, our goal is to study all possible networks consisting of three nodes x_1, x_2, x_3, in which each node can either activate, inhibit or have no influence on any other node in the network. The dynamics of this network of three interacting genes is given by

$$\dot{x}_i = -x_i + r_i(x), \quad i = 1, 2, 3, \tag{4.36}$$

in which each r_i may consist of at most three monotonic activation or inhibition functions. Thereby, only multiplicative combinations are allowed. We furthermore impose the requirement that the resulting network is at least weakly connected, that is, there are no isolated nodes. Concerning the steady state requirements, we study the following two conceptually different mechanisms.

S1) x_1 corresponds to a progenitor factor, maintaining the progenitor state. x_1 is present at high concentration in the progenitor state $\mathbf{x}^{(A)}$, and at low concentration in the differentiated states $\mathbf{x}^{(B)}, \mathbf{x}^{(C)}$. It has to be downregulated in order to achieve cell differentiation. Therefore, the three required stable steady states can by characterized by

$$\mathbf{x}^{(A)} := \mathcal{I}^{x_1,h} \times \mathcal{I}^{x_2,l} \times \mathcal{I}^{x_3,l}$$
$$\mathbf{x}^{(B)} := \mathcal{I}^{x_1,l} \times \mathcal{I}^{x_2,h} \times \mathcal{I}^{x_3,l} \tag{4.37}$$
$$\mathbf{x}^{(C)} := \mathcal{I}^{x_1,l} \times \mathcal{I}^{x_2,l} \times \mathcal{I}^{x_3,h}.$$

S2) x_1 corresponds to a differentiation factor, enabling differentiation. x_1 is present at low concentration in the progenitor state $\mathbf{x}^{(A)}$, and at high concentration in the differentiated states $\mathbf{x}^{(B)}, \mathbf{x}^{(C)}$. It has to be upqregulated in order to achieve cell differentiation. This corresponds to requiring the stable steady states

$$\mathbf{x}^{(A)} := \mathcal{I}^{x_1,l} \times \mathcal{I}^{x_2,l} \times \mathcal{I}^{x_3,l}$$
$$\mathbf{x}^{(B)} := \mathcal{I}^{x_1,h} \times \mathcal{I}^{x_2,h} \times \mathcal{I}^{x_3,l} \qquad (4.38)$$
$$\mathbf{x}^{(C)} := \mathcal{I}^{x_1,h} \times \mathcal{I}^{x_2,l} \times \mathcal{I}^{x_3,h}.$$

In this, we define all low intervals as $\mathcal{I}^{x_i,l} = [0, 0.3]$ and all high intervals as $\mathcal{I}^{x_i,h} = [0.7, 1]$.

Structural properties

For all possible interaction structures and specifications S1 and S2, we have computed the robustness measure $\mathcal{R}$ according to Problem 4.28. As a first result, we have found that there are only a few possibilities for the structure of a production term $r_i(x)$, meaning that the network can only be composed of a limited number of building blocks. This result is illustrated and explained in Figure 4.7. A very similar result can be given for specification S2, which is illustrated in the same figure. As there are less building blocks for hypothesis S1, the number of networks satisfying this hypothesis is smaller. These networks also show a tendency toward negative, that is, toward inhibiting interactions and hence inconsistent loops. The number of networks that can satisfy hypothesis S2 is higher. As can be seen from the building blocks, these networks have about equal number of positive and negative entries, thus reducing the amount of inconsistent loops in favor of consistent loops. In this context, it is also interesting to note that a differentiation module for mesenchymal stem cell differentiation based on literature data, and analyzed in detail in Schittler *et al.* (2010), satisfies specification S1 and can indeed be composed from the building blocks presented in Figure 4.7.

Robustness properties

An interesting observation can be made by examining how the robustness value of an interaction structure is distributed for distinct structural properties such as the number of interconnections or the difference between the number of activating and inhibiting interconnections. Figure 4.8 shows that there is a negative correlation of $\mathcal{R}$ with increasing nonzero entries, and a positive correlation of $\mathcal{R}$ with the abundance of activating interconnections. Both observations are true independent of the specific specification S1 or S2.

Inspired by Kwon & Cho (2008), we next investigate whether there is a relationship between the maximal number of regulators of a node in the network (that is, the maximal indegree) and its robustness value. Kwon & Cho (2008) have observed that perturbations on nodes involved in large or small number of feedback loops have a large or small impact on the overall network robustness. Leclerc (2008) has introduced the so-called gross average cost of perturbation as a robustness measure that explicitly takes network complexity, that is, the number of interconnections in network, into account. The result of this study was that sparser networks outperform densely connected networks and might thus be preferred during evolution. Finally, also the well-studied gene regulation networks from Escherichia coli, Arabidopsis, or Drosophila show a relatively small connectivity of only 1.5–2 regulating factors per gene in average (Leclerc, 2008).

(S1)-models, every node x_i, $i \in \{1,2,3\}$

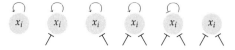

(S2)-models, node x_1

(S2)-models, nodes x_j, $j \in \{2,3\}$

Figure 4.7.: Complete enumeration of all "building blocks". Networks assembled from these building blocks are able to reproduce the respective steady state specification. (Upper row:) Links entering x_i from bottom left and right represent links from nodes x_j, $j \neq i$ and x_k, $i \neq k \neq j$. (Middle row:) Links entering x_1 from below left and right represent links from nodes x_2 and x_3. (Lower row:) Links entering from top right represent links from x_1. Links entering from right represent links from x_k, $k \in \{2,3\}$, $k \neq j$. Thereby, pointed arrows denote activating, and blunt arrows denote inhibiting interactions.

For our example, it should therefore be expected that sparser networks have a larger robustness value than dense structures. Figure 4.9 shows that there is indeed a such correlation. Independent of the specification S1 or S2, networks with a smaller maximal indegree tend to have a higher robustness value. A closer inspection of the respective optimization problems shows that the activation or inhibition functions of the production terms $r_i(x)$ composed of the largest number of individual monotonic functions are the limiting functions for the robustness value. Formulated differently, the more nodes simultaneously influence some other node in the network, the finer these functions need to be tuned in order to satisfy the respective network specification. This observation is even independent of the specific steady state pattern S1 or S2.

In summary, the application of the method for steady state robustness analysis and model discrimination reveals necessary conditions on the interaction structure, such that the minimal three-node models can exhibit a desired steady state pattern. Furthermore, it proposes that these networks should be sparse, which is in good agreement with findings reported previously in the literature and supports the definition of our robustness measure.

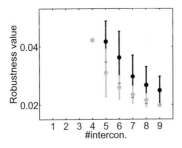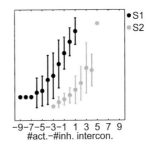

Figure 4.8.: The distribution of the robustness value versus the overall number of interconnections in the network (#intercon.) and versus the difference of activating minus inhibiting interconnections (#act.-#inh. intercon.) in the network. The figures show a negative correlation of the robustness value with the number of interconnections and a positive correlation of the robustness value with the abundance of activating interconnections.

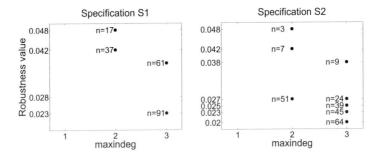

Figure 4.9.: The robustness value $\mathcal{R}$ versus the maximal indegree (maxindeg) for all models. For each data point, there are n models which take these values.

4.6. Summary and discussion

In this chapter, we have addressed the problems of model validation and discrimination with respect to a desired multistable behavior. The focus on multistability is justified by its importance for gene regulation networks, especially for those involved in differentiation and pattern formation. A modeling framework based on ordinary differential equations has been extended, which is especially tailored to describe multistable gene regulation networks. It allows to address both, qualitative and quantitative aspects about the system itself and the measurements.

Based on this framework, we have developed an algorithm to decide in a "yes" or "no" fashion, whether a hypothetical interaction structure is at least in principle capable to reproduce a desired steady state pattern. It is intended as a first step in the modeling process to quickly sort out unsuitable model structures and only refine those in later stages of the modeling process, which have passed this validation step. Its benefits have been demonstrated in a hypothetical scenario in the context of the lactose utilization network.

Building on this first contribution, a second method has been developed. It requires more quantitative information such as numeric values for the concentrations of the individual species at the steady states, but also allows for stronger conclusions about the systems behavior. It cannot only be used to decide, whether the system can reproduce the observed multistable behavior, but also, how well it can do so. For this second aspect, we have defined a novel robustness measure for the interaction structure itself. Unlike existing approaches in the literature, it does not evaluate the robustness of a given parameterized nominal system, but considers all possible realizations of the involved activation or inhibition functions. The robustness measure of the interaction structure has been defined to be the robustness measure of its best possible realization. The reasoning behind this approach was the common claim, that the natural evolution and selection process has lead to systems, which can realize their function, which was multistability in this chapter, in a maximally robust way. Furthermore, we have devised an efficient formulation as a convex optimization problem to compute this robustness measure for the two important subclasses of systems, in which all monotonic functions are either always combined in an additive or a multiplicative way.

Similar to the first method, also this second approach is intended to support modelers in early stages of the modeling process. We have applied the method to uncover structural properties of minimal three-node networks, which allow these networks to act as robust decision modules in the progress of stem cell differentiation. Thereby, we have found that sparsity is the key to robustness in this scenario, which is in good agreement with previous findings in the literature. By inspecting the results of the respective optimization problems and identifying the constraints active at the optimal solution, one can furthermore identify the most fragile interconnections in a system.

In some respects, this approach also resembles flux-balance and energy-balance analysis of metabolic networks (Edwards *et al.*, 2001; Beard *et al.*, 2002). For these methods, too, the structure or stoichiometry of a network is assumed to be known, but only qualitative constraints on the reaction kinetics are imposed. Then, these approaches also employ (convex) optimization to maximize certain performance criteria. While this objective was the maximization of the robustness measure in our case, flux-balance or energy-balance typically aim to maximize energy production, the growth rate or other biologically important factors. However, there is no direct relation to the methods developed in this chapter because of the inherent differences between signaling and metabolic networks. In conclusion, with the two methods for model validation and discrimination, all goals formulated in Chapter 4.1.3 have been achieved.

5. Control of Boolean models of gene regulation networks

In this chapter, the problem of guiding Boolean systems of gene regulation networks toward a desired attractor is studied. The problem is motivated and formalized in Section 5.1. Also related results from the literature and a brief introduction to discrete event systems is given there. In Section 5.2, the control problem is transformed into a state attraction problem for discrete event systems. Also extensions concerning limitations on the applicability of inputs and observations are discussed. In Section 5.3, the new methodology is applied to an example system. Section 5.4 gives a summary and conclusions.

The chapter uses the methodology developed by Schmidt & Breindl (2014), which is however not part of this thesis.

5.1. Introduction and problem statement

The problem of designing control or intervention strategies for gene regulation networks arises in many applications. For the design of artificial networks in a synthetic biology context, the question of how different genes and their products should interact in order to achieve a desired network behavior, is most fundamental (Purnick & Weiss, 2009). But also outside the field of synthetic biology, the control problem is highly relevant. Numerous studies are concerned with the problem of finding strategies to influence the behavior of regulatory, signaling or metabolic networks in order to ease diseases, guide cell differentiation toward desired types, or to optimize the production of bacterial metabolites (Shmulevich *et al.*, 2002b; Baldissera & Cury, 2012).

In this chapter we will now turn to the problem of manipulating and controlling gene regulation networks. We will again consider qualitative models of gene regulation networks which evolve in discrete time on a discrete state space, using the same reasoning as before that these models are complex enough to reproduce the real network behavior in sufficient detail. The control objective can be derived from Section 4.1. There, we have emphasized the importance of attractors for the phenotypes resulting from the network dynamics. Therefore, the goal of a control strategy will now be to steer the system to a desired attractor. To give a few examples, note that the problems of guiding stem cell differentiation to desired cell states or to enforce Apoptosis in cancer cells can be cast as such a state convergence problem (Foster *et al.*, 2009; Chaves *et al.*, 2009; Calzone *et al.*, 2010).

Unfortunately, the achievement of the control goal in a realistic biological setting is severely hampered by fundamental limitations concerning the applicability of external stimuli and the possibility to observe the reaction of the network to these stimuli. Usually, only few nodes in the network can be influenced, and the control actions that can be exerted on those nodes are limited. Consider for example the cell-fate decision model in Calzone *et al.* (2010), which will be studied in detail in Section 5.3. This study provides a high-level model of the genes and signaling pathways which become active upon stimulation by TNF or FAS ligand and decide if this stimulation leads to Apoptosis, to a non-apoptotic cell death, or to survival by activation of the NFκB pathway. It is obvious that the natural inputs to the system, TNF and FAS ligand, can not be added or removed arbitrarily often, as washing out the ligands is experimentally expensive. As another example, consider the case that the expression of a certain gene is amplified by injecting a plasmid ring into the cell, which contains the respective gene sequence. Then, this gene should be considered as "on" in our qualitative framework, and, unless it requires highly specific transcription factors, it is difficult to turn it "off" again.

The second barrier for the development of effective control strategies is the fact that the internal state of the system, that is, the abundance of the species, is hard to observe, at least if we want to obtain real time and non-destructive measurements. Often, one has to rely on indirect observations, such that this internal state needs to be deduced from phenotypes. One measurement technique, which allows to obtain real time and non-destructive observations, is the tagging of proteins by fluorescent markers (see Chapter 2.3). The abundance of these proteins can then be detected by microscopy. However, the number of proteins which can be tagged and distinguished by this method is limited and does not exceed two or three in the studies known to the author. Therefore, a method to design control strategies for gene networks should not assume that the full state of the model is known.

The goal of this chapter is to find a solution to the control problem under the limitations as just described. In difference to the previous chapters, we will however not develop completely new methods, but make use of tools that already exists in the literature, albeit formulated in different frameworks: in a discrete event systems (DES) setting, the methods developed for state attraction (Brave & Heymann, 1990), and state attraction under partial observation (Schmidt & Breindl, 2014) already provide all formulations required to approach this problem. The main contribution of this chapter is therefore the transformation of the control problem into a discrete event systems setting. We furthermore want to demonstrate that the DES framework is a reasonable choice to approach the control problem under these severe limitations, and allows to develop successful control strategies whenever they exist.

5.1.1. Problem formulation

From the above explanations we will now deduce the main problem of this chapter. Again, we will use a Boolean system description. However, in contrast to the previous chapter, we now include control inputs, such that the model equations take the form

$$x_i(t+1) = B_i(x(t), u(t)), \quad i = 1, \ldots, n. \tag{5.1}$$

As in Chapter 3, $x \in \{0,1\}^n$ represents the expression state of the n genes in the network, $u \in \{0,1\}^p$ denotes the discrete vector of p control inputs, and the functions $B_i : \{0,1\}^{n+p} \to \{0,1\}$ are known update functions. We may consider synchronous or asynchronous updates to generate trajectories from (5.1). As asynchronous updates are more realistic and lead to a more complex system behavior, and thus to a more complex control problem, the focus will be on asynchronous updates. Let us furthermore define a set X^o which contains all state variables whose value can be observed. As explained in the introduction, we typically have $p < n$, and therefore $X^o \subset X := \{x_1, \ldots, x_n\}$. With slight abuse of notation, we will also use X^o as a vector containing all observable states.

The control problem can then be described as follows. Assume a set $\mathcal{X}_a$ is given, which contains one or several attractors of System (5.1) for some fixed inputs. That is, $\mathcal{X}_a$ is invariant under asynchronous updates of (5.1) with $u(t)$ set to some fixed value $\hat{u} \in \{0,1\}^p$. Assume furthermore that a set $\mathcal{X}_0$ containing one or several initial configurations for the state x is given. We assume throughout this chapter that $u = 0$ for each such initial configuration $(x, u) \in \mathcal{X}_0$. The goal is then to compute a control strategy Π, such that the system is steered to $\mathcal{X}_a$ whenever it starts in a state from $\mathcal{X}_0$. Thereby, it is not important which state $x_a \in \mathcal{X}_a$ is reached if $\mathcal{X}_a$ contains several attractors.

Concerning the limitations described above, the control strategy is only allowed to use information about the evolution of the observable states X^o, that is, we can consider Π as a function $\Pi(X^o(t), X^o(t-1), \ldots)$. This means, we allow Π to not only depend on the present values of the observable states, but also on their past. Finally, we want to be able to decide about the existence of such a control strategy Π, and compute Π whenever it exists. The controlled system $x_i(t+1) = B_i(x(t), \Pi_i(X^o(t), X^o(t-1), \ldots))$ should then be guaranteed to reach one of the attractors in $\mathcal{X}_a$ in a finite number of updates and remain there indefinitely.

5.1.2. Established approaches

Several approaches toward the control of discrete models of gene regulation networks have been presented before. In this context, a framework, which has received considerable attention in the literature, is the class of probabilistic Boolean networks (PBNs), introduced by Shmulevich *et al.* (2002a). PBNs are a superclass of Boolean networks in which there is not only one update function for each node, but a whole set. At each discrete point in time, one of these update functions is chosen randomly to compute the successor state. Trajectories can be generated using synchronous or asynchronous updates. Control strategies for PBNs have been proposed in a number of publications. In Shmulevich *et al.* (2002b), PBNs are used together with synchronous updates. As reasonable strategies, control actions are suggested which maximize the probability of reaching a certain state, or minimize the expected time until a certain state is reached. In Datta *et al.* (2003), these problems are formulated as finite-horizon optimal control problem minimizing a given performance index. It is shown that this problem can be solved by dynamic programming. This was followed by extensions such as a model for the switching of the update functions (Pal *et al.*, 2005), approximations of the control problem for large systems (Ng *et al.*, 2006), and papers summarizing these works (Datta *et al.*, 2006). There are also more recent works, such as Liu (2012), which aim to refine the original problem from Shmulevich *et al.* (2002b). While all the references mentioned so far assume that the full state of the system can be observed, the case of only partial state information is considered and related to the theory of controlled Markov chains under imperfect state information by Datta *et al.* (2004). All these methods yield probabilistic results and assume that the inputs can be set arbitrarily.

Also more conceptual studies not directly related to the field of modeling and controlling gene regulation networks have been performed, also considering discrete models as in Equation (5.1). Akutsu *et al.* (2007) have shown that the control problem for Boolean networks is NP hard. Cheng *et al.* (2011) and Li & Sun (2011) transfer controllability and observability conditions known from control theory for systems of differential equations to the Boolean and discrete case. Characterizations are then given based on a so-called semi-tensor product. However, these formulations yield extremely large expressions and matrices which can, according to own performance tests, not even be handled for medium sized systems of ten nodes.

With respect to other discrete model frameworks, gene regulation networks have also been modeled as discrete event systems or Petri nets before. It was shown in Steggles *et al.* (2006) how a Boolean model as in (5.1), yet without inputs, can be transferred into a Petri Net. In Chen & Weng (2009) and Chen & Weng (2011), automata models for gene regulation networks were derived from a Boolean description. This approach however requires a large amount of smaller automata, as every node in the network as well as each update function is modeled by an individual automaton. The overall network behavior then results from the parallel composition of these automata. The work presented by Baldissera & Cury (2012) builds upon this framework and discusses the applicability of ideas based on supervisory control theory for discrete event systems to achieve a desired network behavior. The work is however very conceptual and gives no formal design methods or algorithms.

As we will also derive a discrete event systems model, especially the work of Ramadge and Wonham (Ramadge & Wonham, 1987) is to be mentioned, which has laid the basis for the supervisory theory of discrete event systems. The algorithms to solve the state attraction problem in this framework have been presented by Kumar *et al.* (1993) for the case of full observation, and by Schmidt & Breindl (2014) for the case of partial observation.

5.1.3. Challenges and goals

As seen in the previous section, there is no approach in the literature which can address all of the following requirements.

- Computation of a control strategy, whenever it exists, which can guarantee convergence to the desired attractor.

- Consideration of limitations of when and how inputs can be applied.

- Only partial observation.

However, as outlined above, solutions to handle these requirements are available in the field of discrete event systems. In this chapter, we will therefore transform the control problem into a state attraction problem for discrete event systems. This reformulated problem can then be treated with existing solution algorithms. Some extensions to these existing methods are however necessary in order to be able to address all of the above requirements. The individual goals are therefore described next in more detail.

To begin, we need to generate an automaton P, which can exactly reproduce all possible trajectories of System (5.1), or equivalent representations thereof. The limitations of how and when control inputs can be applied will be modeled by a second automaton, K, such that the parallel composition $G := P||K$ exactly reproduces all possible behaviors of the controlled network. To demonstrate this approach, we will consider the exemplary scenario that each input can only be altered one time, and that the value of u can only be modified after a change in the expression of the observable states has been observed. These restrictions are in good agreement with experimental practice. Yet, one can easily imagine other scenarios and their incorporation into the solution approach can be done in much the same way.

Finally, the C++ library libFAUDES (Moor *et al.*, 2008) already offers a wide collection of tools to represent and manipulate discrete event systems, including an implementation of the algorithm for state attraction under full event observation form Kumar *et al.* (1993). Therefore, the generation of the automata P, K and G should also be implemented in C++ using this library. Furthermore, the algorithm for state convergence under partial observation form Schmidt & Breindl (2014) needs to be implemented, thus extending the libFAUDES library. The main goals of this chapter can therefore be summarized as follows.

- Find an automaton representation for the controlled system, respecting the limitations on when and how control inputs can be set.

- Application of the methods for state attraction under full or partial observation to compute, whenever they exist, successful control strategies.

5.1.4. Some fundamentals on discrete event systems

The most basic definitions, concepts and notations of discrete events systems, as required for the understanding of this chapter, are now briefly recalled. An alphabet is a set of distinct symbols or events $\Sigma = \{\sigma_1, \sigma_2, \ldots, \sigma_m\}$. Any sequence of events forms a string, and ϵ denotes the empty string. A language L is a set of finite strings, $L \subseteq \Sigma^*$, and Σ^* contains all possible finite strings consisting of symbols from Σ.

An important operation on languages, which is also needed for this chapter, is the natural projection $p_a : \Sigma^* \to \Sigma_a^*$, for $\Sigma_a \subseteq \Sigma$. It is defined iteratively by: (1) $p_a(\epsilon) := \epsilon$ and (2) for $s \in \Sigma^*$, $\sigma \in \Sigma$, $p_a(s\sigma) = p_a(s)\sigma$ if $\sigma \in \Sigma_a$, and $p_a(s)\sigma = p_a(s)$ otherwise. In words, this means that the natural projection p_a erases all symbols from a string, which are not in Σ_a.

A discrete event system is a finite state automaton $G = (Z, \Sigma, \delta, Z_m, Z_0)$, in which Z is the finite set of states, Σ is the finite alphabet of events, δ is the partial transition function $\delta : Z \times \Sigma \to Z$, $Z_m \subseteq Z$ is the set of marked states, that is, states with a special meaning, and $Z_0 \subseteq Z$ is the set of initial states. Assume now that G is in state $z_a \in Z$, and denote by $\Lambda(z_a)$ the set of events that are possible at state z_a. Then, any transition $(z_a, \sigma) \to z_b$ can occur for which $\sigma \in \Lambda(z_a)$.

The transition function also establishes the connection between languages and automata by considering all feasible sequences of transitions. The generated language of G is the set of all strings that can occur along the execution of transitions in G, starting at an initial state from Z_0. The marked language is the set of all strings σ, such that a marked state is reached in G after the occurrence of σ. The most important operation on automata for this chapter is the synchronous product of two automata G_1 and G_2. It models the synchronized parallel execution of these two automata, and is defined by $G_1||G_2 := G_{1,2} = (Z_1 \times Z_2, \Sigma_1 \cup \Sigma_2, \delta_{1||2}, Z_{0,1} \times Z_{0,2}, Z_{m,1} \times Z_{m,2})$, with

$$\delta_{1||2}((z_1, z_2), \sigma) := \begin{cases} (\delta_1(z_1, \sigma), \delta_2(z_2, \sigma)) & \text{if} \quad \sigma \in \Lambda_1(z_1) \cap \Lambda_2(z_2) \\ (\delta_1(z_1, \sigma), z_2) & \text{if} \quad \sigma \in \Lambda_1(z_1) - \Sigma_2 \\ (z_1, \delta_2(z_2, \sigma)) & \text{if} \quad \sigma \in \Lambda_2(z_2) - \Sigma_1 \\ \text{undefined} & \text{else.} \end{cases}$$

Note that, at a state (z_1, z_2), a shared event, that is, an event $\sigma \in \Sigma_1 \cap \Sigma_2$, can only be executed if it is in the active event set of the corresponding two states x_1 and x_2 of G_1 and G_2. Thus, two automata are synchronized on the shared events. All other events are possible whenever they are generated by G_1 or G_2. A state of the resulting automaton is only marked if the corresponding states of G_1 and G_2 are both marked.

For the purpose of controlling the behavior of G, the alphabet Σ is partitioned into the set of controllable events, which can be prevented from occurring by a supervisor, and the set of uncontrollable events. Furthermore, Σ can be partitioned into the set of observable events, whose occurrence can be observed by a supervisor, and the set of unobservable events, whose occurrence is not visible.

Finally, the state attraction problem for a discrete event systems $G = (Z, \Sigma, \delta, Z_m, Z_0)$ is to find a supervisor which guarantees that a predefined subset of X' is reached in a limited number of steps from every initial state in X_0, and that the system remains in X' thereafter. The problem has been solved in Brave & Heymann (1990) for the case that all events can be observed, and for the case of only partial observation in (Schmidt & Breindl, 2014). As the results and the algorithmic solutions from these references are not necessary for the understanding of this chapter, they are summarized in Appendix C.

5.2. Representation of Boolean networks as a discrete event system

We now derive an automaton representation of all possible controlled behaviors of the Boolean network. As already outlined, we proceed in two steps. First, an automaton is constructed that can reproduce all possible controlled behaviors without respecting limitations on the applicable inputs. A second automaton will then model the limitations on the control inputs.

The first automaton, $P = (Z, \Sigma, \delta, Z_m, Z_0)$, is the state transition graph of (5.1), in which we attach certain interpretations to the states and arcs and define observability and controllability properties. Its construction is described in Algorithm 5.1. As with state transitions graphs, each state z in the automaton P represents a unique Boolean state $(x, u) \in \{0, 1\}^{n+p}$, which is denoted by $(x, u) \sim z$. This can for example be implemented algorithmically by computing a decimal equivalent to (x, u), such that z takes decimal values from 0 to $2^{n+p} - 1$, and which yields a bijective mapping between the Boolean states (x, u) of System (5.1) and the states z of P. Assuming such a bijective mapping, the automaton state z and the corresponding Boolean state (x, u) will be used interchangeably.

The alphabet Σ of P consists of three types of events. In line 3 of Algorithm 5.1, we define the events which correspond to setting the control input u to some value $u \in \{0, 1\}^p$. These events are observable and controllable. To characterize the asynchronous updates of the original system, the following two classes of events are introduced. First, events which corresponds to a change in an observable state $x_i \in X^o$ are defined in line 5. As we can observe, but not prevent the occurrence of the corresponding transitions, these events are observable and uncontrollable. Second, the events representing transitions of states $x_i \in X \setminus X^o$ are introduced in line 7. As these transitions can not be observed, these events are uncontrollable and unobservable. As with asynchronous updates only one state x_i can alter its value at a time, these events are sufficient to characterize all possible transitions between two states.

From an algorithmic perspective, the use of a stack (line 8) allows to only create those states and transitions, which are accessible from the initial configurations in $\mathcal{X}_0$, such that Z may contain significantly less than 2^{n+p} states. All states of P corresponding to attractor states in $\mathcal{X}_a$ are furthermore added to the set of marked states. Note that these states are characterized by having only self-loops of events e-u for $u \in \{0,1\}^p$.

The relation between P and the original system

Any sequence of events that can be generated by the automaton P corresponds to a possible trajectory of System (5.1) in the sense that the sequence of states z visited by such an execution of events corresponds to the sequence of Boolean states $(x,u) \sim z$ along that trajectory. Thereby, we model the application of a new control input, that is, the change of u from $\hat{u}$ to $\bar{u}$, by a sequence of Boolean states $(x(t), \hat{u}(t))$, $(x(t+1), \bar{u}(t+1))$, in which x remains constant. Thus, setting a new control input requires one discrete time step, which is slightly unconventional but guarantees the equivalence between executions of P and trajectories of System (5.1). It however imposes no restrictions, as there is no real time interval attached to a discrete time step.

Assume now that an execution of P has started in z_0, and that we have observed a string $\sigma = o_1 o_2 \ldots o_m$ of observable events. Then, if we know the Boolean values of all the observable states $x_j \in X^o$ at $(x_0, u_0) \sim z_0$, we can reconstruct the Boolean values of these observable states after the observation of σ in a unique way. However, all other states $x_j \in X \setminus X^o$ can not be reconstructed uniquely, and, because of the randomness of the updates, there might be many states that can be reached by the same observable strings. Formally, this means that there might exist two different states $z_1 = \delta(z_0, s_1)$, $z_2 = \delta(z_0, s_2)$ with $p(s_1) = p(s_2) = \sigma$ but $s_1 \neq s_2$.

As information about the current state z of P can only be deduced from the observable events, we will later extend this automaton to make the information about the Boolean state of the observable states $x_j \in X^o$ at an initial configuration z_0 visible to a controller. Before that, we will however turn to the development of a description of the limitations on the control inputs.

5.2.1. Modeling of control and observation scenarios

As outlined in Section 5.1.1, we want to consider the exemplary scenario that new inputs can only be set immediately after the observation of a transition in an observable state, and that each input can only be modified one time. Recall that we have assumed that $u = 0$ holds at all initial configurations. Algorithm 5.2 now presents a way, how this requirement can be translated into an automation representation. The resulting

---

Algorithm 5.1: Creation of the automaton P.

```
// preparations
```
1 create an empty automaton $P = (Z, \Sigma, \delta, Z_m, Z_0)$
2 **foreach** $u \in \{0,1\}^p$ **do**
3   add the observable and controllable event e-u to Σ
4 **foreach** *observable state* $x_i \in X^o$ **do**
5   add the the observable and uncontrollable events x_i-up and x_i-down to Σ
6 **foreach** *unobservable state* $x_i \in X \backslash X^o$ **do**
7   add the the unobservable and uncontrollable events x_i-up and x_i-down to Σ

```
// construction of P
```
8 create an empty stack todoStates
9 **foreach** *initial configuration* $(x_0, u_0 = 0) \in \mathcal{X}_0$ **do**
10   create state $z \sim (x_0, u_0)$ and add it to Z and Z_0
11   push z onto todoStates
12 **while** todoStates *is not empty* **do**
13   pop off top element z of todoStates
14   call computeSuccessors(z,todoStates,P)
15   add the states $z \in Z$ corresponding to attractor states in $\mathcal{X}_a$ to Z_m

```
// helper function
```
16 **Function** computeSuccessors(z,todoStates,P)
17   compute Boolean state $(x, u) \sim z$
18   **foreach** $\hat{u} \in \{0,1\}^p$ **do**
19   **if** $\hat{z} \sim (x, \hat{u})$ *does not exist in* Z **then**
20   create state $\hat{z}$ in Z and push it onto todoStates
21   set transition $\delta(z, e\text{-}\hat{u}) = \hat{z}$
22   **foreach** *asynchronous successor state* $(\tilde{x}, u)$ *of* (x, u) *according to* (5.1) **do**
23   **if** $\tilde{z} \sim (\tilde{x}, u)$ *does not exist in* Z **then**
24   create $\tilde{z}$ in Z and push it onto todoStates
25   **if** $\tilde{x}_j > x_j$ *for some component j* **then**
26   set transition $\delta(z, x_j\text{-up}) = \tilde{z}$
27   **if** $\tilde{x}_j < x_j$ *for some component j* **then**
28   set transition $\delta(z, x_j\text{-down}) = \tilde{z}$

---

automaton is denoted by $K = (Y, \Sigma, \gamma, Y_m, Y_0)$. Figure 5.1 illustrates the resulting automaton for the case that Σ consists of the observable but uncontrollable events x_1-up and x_1-down, the unobservable and uncontrollable events x_2-up and x_2-down, and the observable and controllable events e-u for $u \in \{0,1\}^2$.

Let us proceed to describe the meaning of the individual states and transitions of K in more detail. Also recall that the final automaton reproducing all controlled executions of System (5.1) under consideration of the input limitations is obtained as parallel composition of P and K, $G := P||K$. As the system might start in an attractor, and as we assume $u = 0$ at all initial configurations, we first enforce the application of some input $u \in \{0,1\}^p$. This is represented by the four transitions $e - u$ leaving the initial state 1 in Figure 5.1, and implemented in lines 2 to 6 of Algorithm 5.2. After that, the system can undergo any number of unobservable transitions while the current state in K is not left. This is realized by the self-loops of all unobservable and uncontrollable events at the current state of K, and implemented in lines 8 and 9. Then, after the occurrence of some observable and uncontrollable event, a new state in K is entered, in which we have to reconsider the control inputs to be applied (lines 11 and 12 of the algorithm). It is now possible to leave the input unchanged, modeled by applying the same input again which is currently set (lines 15 and 16), or to apply any other input, for which at least one more component is set to 1 compared to the current input. All states that are reached after an event $e - u$ are marked states. The effect of the marked states on $G = P||K$ will be discussed in the following section.

Algorithm 5.2: Creation of the scenario automaton K.

---

1 create an automaton $K = (Y, \Sigma, \gamma, Y_m, Y_0)$, which has the same alphabet as P
2 add a new state y_0 and add it to Y and Y_0
3 **foreach** $u \in \{0,1\}^p$ **do**
4 | add a new state y_1 to Y and Y_m
5 | set transition $\gamma(y_0, e\text{-}u) = y_1$
6 | call computeSuccessorsK(y_1,K)

    // helper function
7 **Function** computeSuccessorsK(y, K)
8 | **foreach** *unobservable event* $\sigma \in \Sigma$ **do**
9 | | set transition $\delta(y, \sigma) = y$ in K
10 | create a new state y_1 in Y
11 | **foreach** *observable event* $\sigma \in \Sigma$ **do**
12 | | set transition $\delta(y, \sigma) = y_1$ in K
13 | collect the indices j of all inputs, such that $u_j = 1$ at state z, in a set I_s
14 | compute $I_n = \{1, 2, \ldots, p\} \setminus I_s$
15 | compute u such that $u_j = 1$ exactly if $j \in I_s$
16 | set transition $\delta(y_1, e\text{-}u) = y$ in K
17 | **foreach** *nonempty subset* $J \subseteq I_n$ **do**
18 | | compute u such that $u_j = 1$ exactly if $j \in J \cup I_s$
19 | | add a new state y_2 to Y and Y_m
20 | | set transition $\delta(y_1, e\text{-}u) = y_2$ in K
21 | | recursively call computeSuccessorsK(y_3,K)

5.2.2. The composed automation $G = P||K$

The overall automaton $G = P||K$ is obtained from the synchronous product of P and K. From the construction of P and K and the definition of the synchronous composition,

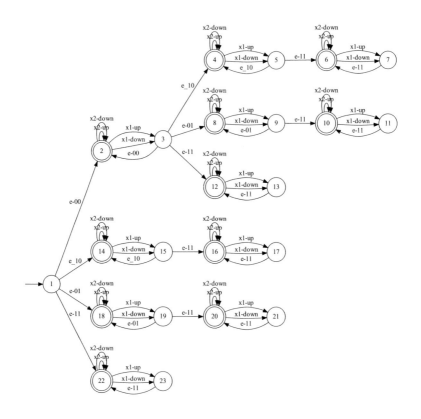

Figure 5.1.: Automaton K modeling the limitations on the application of inputs. In this example, all events *e-u* correspond to setting a control input, the events $x1$-up and $x1$-down correspond to changes in the observable node x_1, and events $x2$-up and $x2$-down correspond to changes in the unobservable node x_2. Marked states are depicted as double circles.

it is clear that G contains all possible controlled sequences of System (5.1), in which the application of control inputs is subject to the limitations. A result about the marked states of P can be stated immediately.

Lemma 5.1. *In G, exactly those states $q = (z, y)$ are marked, for which z corresponds to an attractor state in $\mathcal{X}_a$.*

The advantage of this lemma is that the identification of the states of G corresponding to an attractor state are already identified during the computation of the synchronous product, and need not be determined separately.

Finally, we need to discuss one further aspect that has already been mentioned at the end of Section 5.2. Consider the case that there are several possible initial configurations in $\mathcal{X}_0$, which differ by the value of some of the observable states from X^o. As information about the state of the system can only be deduced from the occurrence of events, a supervisor for G can not distinguish these states. In order to also allow a supervisor to distinguish these states, we suggest the following postprocessing step for all initial states of G. Let $q_o = (z_0, y_0)$ be an initial state and let $(x_0, 0) \sim z_0$ be the Boolean initial configuration corresponding to z_0. Then, replace the state q_0 by a linear sequence of states and events along which x_j-up occurs exactly once for each observable state x_j, which has the Boolean 1 at x_0. The last state along this sequence then inherits all transitions from q_0. By this construction, available information about the initial configuration can be deduced from the occurrence of events. This postprocessing step is illustrated in Figure 5.2.

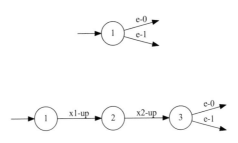

Figure 5.2.: Postprocessing of an initial state of G. In this example, the observable states x_1 and x_2 have Boolean value 1 at the corresponding Boolean initial state. Therefore, initial state 1 is replaced by the linear sequence of states and events shown below.

With this postprocessing, any information that can be obtained about the state (x, u) of the Boolean system from the observable states X^o can also be obtained about the corresponding state q of G from the string of events leading to that state. The following result can therefore be stated.

Proposition 5.2. *There exists a control strategy $\Pi(X^o(t), X^o(t-1), \ldots)$ for the Boolean system, if and only if the state attraction problem for G has a solution.*

In summary, the proposed procedure to construct two different automata P and K has the advantage, that their generation is conceptually easier than the direct construction of G. This direct construction would require to distinguish various cases of how a certain Boolean state is reached, such as for example, if it is reached by an observable or by an unobservable event. Moreover, if a different scenario is considered, only K needs to be replaced in a modular fashion, while all other algorithms remain the same.

5.2.3. Controllability and observability properties

As described above, the actual design of a control strategy Π as requested in Section 5.1.1, realized in the form of a supervisor automaton S, has been presented in Kumar *et al.* (1993) and Schmidt & Breindl (2014), and we refer to Appendix C for a brief summary. Here, we want to discuss two relevant question concerning the states to be observed and the inputs to be applied in order to achieve the control objective. For biological applications, it is preferable to keep the set X^o of states, which need to be observed, as small as possible, and to use as few inputs as possible. Thereby, an unused input is constantly set to zero. The following lemmas are straight-forward and can help to find a minimal set X^o, and a minimal set of used inputs.

Lemma 5.3. *Given a set of observable states X^o. Assume that there exists no control strategy $\Pi(X^o(t), X^o(t-1), \ldots)$. Then, for any set $\bar{X}^o \subseteq X^o$, there also exists no control strategy $\Pi(\bar{X}^o(t), \bar{X}^o(t-1), \ldots)$.*

Proof. If there exists a control strategy using information about the states in $\bar{X}^o$, then this is also a control strategy for the case that information about the states in X^o is available. □

Lemma 5.4. *Given a set of observable states X^o and a set of inputs U. Assume there exists no control strategy $\Pi(X^o(t), X^o(t-1), \ldots)$ that only uses the inputs $u_j \in U$. Then, there also exists no control strategy $\Pi(X^o(t), X^o(t-1), \ldots)$ which only uses the inputs in $\bar{U} \subset U$.*

Proof. If there exists a control strategy using only the inputs in $\bar{U}$, then this is also a control strategy using only inputs in U. □

With these results, the following procedure is reasonable to find minimal sets of events to be observed such that a control strategy can be found. Thereby, we mean with minimal that no further state can be removed from the set of observable events such that a control strategy exists. Equivalently, no input can be removed from a minimal set of inputs such that a control strategy can be found. To find such a minimal set X^o, first check if a control strategy exists for the case of full observation. If it exists, states can successively be removed from X^o. Doing this in a systematic way will uncover the, or, all minimal sets X^o such that a control strategy exists. For a fixed set of observable events, the same procedure is suggested to find all, or, the minimal sets of inputs.

5.3. Application example

The methodology developed in this chapter is now applied to a Boolean model from the literature. We will show that the method is not only suited to compute control strategies to achieve a desired system behavior, but also allows to gain a deeper understanding of the system by analyzing its possible behaviors. The model we want to control and analyze has been presented by Calzone *et al.* (2010), and describes the gene regulation and signaling components involved in a decision process about death or survival of the cell. The interaction graph and the update functions are shown in Figure 5.3. The external inputs to the model are TNF and FASL, which represent the stimulation of the system by tumor necrosis factor and Fas ligand. The intracellular processes triggered by these inputs lead either to the survival of the cell, caused by activation of the NFκB pathway, to the non-apoptotic or necrotic cell death, caused by depletion of ATP, or to the apoptotic cell death, caused by activation of CASP3. Each of these three cell fates as well as the initial living state of the cell correspond to a number of point attractors of the model. Due to the randomness introduced by the asynchronous updates, the model can reach each of these three cell-fate attractors from the living initial state upon constant stimulation by either of the two inputs. Also variation of the duration of the applied inputs does not change this situation. For further details on the model dynamics and a deeper discussion of the biological background, we refer to Calzone *et al.* (2010). Using the formalisms presented in this chapter, we now want to study if and how the system can be steered to the apoptotic attractor, avoiding the other attractors. This is motivated by the fact that, in many types of cancer, the apoptotic cell state is impaired. Being able to enforce apoptosis is therefore an important goal in cancer research.

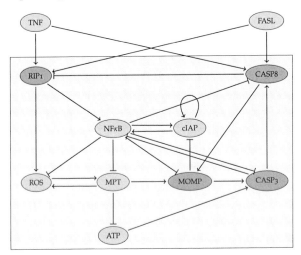

$$B_{RIP1} = \neg CASP8 \wedge (TNF \vee FASL) \qquad B_{NF\kappa B} = (cIAP \wedge RIP1) \wedge \neg CASP3$$
$$B_{CASP8} = (TNF \vee FASL \vee CASP3) \wedge \neg NF\kappa B \qquad B_{cIAP} = (NF\kappa B \vee cIAP) \wedge \neg MOMP$$
$$B_{ATP} = \neg MPT \qquad B_{CASP3} = ATP \wedge MOMP \wedge \neg NF\kappa B$$
$$B_{ROS} = \neg NF\kappa B \wedge (RIP1 \vee MPT) \qquad B_{MOMP} = MPT \vee (C8 \wedge \neg NF\kappa B)$$
$$B_{MPT} = ROS \wedge \neg NF\kappa B$$

Figure 5.3.: Interaction graph and update rules of the Boolean cell-fate decision model from Calzone *et al.* (2010).

In the model, the initial living state is characterized by ATP = cIAP = 1, and all other nodes and inputs having the Boolean value 0. For the apoptotic state, there are four point attractors, each characterized by RIP1 = NFκB = ROS = MPT = cIAP = 0, CASP8 = ATP = MOMP = CASP3 = 1. The inputs however can take all four values $(TNF, FASL) \in \{0,1\}^2$. Thus, for our setup, $\mathcal{X}_0$ consists of only one, and $\mathcal{X}_a$ consists of four point attractors.

General properties of the controlled behavior

Let us first study the question if it is possible at all to steer the system from $\mathcal{X}_a$ to $\mathcal{X}_0$. Thereby, we assume that all states can be measured and that there are no input limitations. To this end the automaton P is constructed according to Algorithm 5.1, which comprises 1376 states, 22 events, and 10768 transitions. Application of the state attraction algorithm for the case of full observability then shows that there exists no such control strategy. This means that no control strategy can prevent the system from possibly ending up in one of the other two attractors, corresponding to the necrotic or the NFκB-survival state. While the simulation study in Calzone *et al.* (2010) already suggested this result, the existence of a control strategy guaranteeing convergence to $\mathcal{X}$ could not be excluded so far. Only our analysis delivers the proof, showing that the control inputs TNF and FASL are not chosen well enough to limit the randomness of asynchronous updates sufficiently.

Extensions of the model

In the light of this result, it is therefore reasonable to ask, if modifications of the system could lead to the existence of such a control strategy. Therefore, we next study several minimal extensions of the model, each introducing one additional input u_a. Thereby, we replace the update rule B_{x_j} of a node x_j by $B_{x_j} \wedge \neg u_a$, if x_j has the Boolean value 0 at the apoptotic attractor states, and by $B_{x_j} \vee u_a$, if it has the Boolean value 1 at the attractor state. This corresponds to the additional ability to down- or up-regulate the expression of x_j, and leaves the attractor $\mathcal{X}_a$ qualitatively unchanged. The apoptotic attractor of each modified systems contains now the eight states (x, u), with x as before, and $u = (\text{TNF}, \text{FASL}, u_a) \in \{0,1\}^3$ instead of $u \in \{0,1\}^2$.

For each possible extended system, the automata P, K, and $G = P||K$ have been constructed as described in this chapter. With the additional input, the respective automata become considerably larger than before. For example, when u_a is applied to RIP1, the automaton G comprises 9629 states, 26 events, and 26871 transitions. In the other cases, the resulting automaton sizes are in the same order of magnitude. The algorithm for state attraction under full observation has then been applied to P and K, that is, it has been tested if there exists a control strategy for the case of no input limitations, and for the case that each input can only be set once. Interestingly, the results are the same. Table 5.1 gives a brief summary of these results.

Thus, with inputs TNF, FASL, and u_a applied to either RIP1, MOMP, CASP8, or CASP3, convergence to $\mathcal{X}_a$ can now be guaranteed. These nodes are highlighted in darker gray in Figure 5.3. Moreover, for every of these three cases, the input u_a alone was sufficient. This can for example be seen in Figure 5.4, where the controlled behaviors with u_a applied to CASP8 are shown. One can see that there is one path from the initial state to a marked state, along which only the control input $e - 01$ (corresponding to setting $u_a = 1$, see the caption for more details) is applied. Thus, direct control of either of the four species RIP1, MOMP, CASP8, or CASP3, is more effective than the two natural inputs TNF and FASL, as this either directly activates the apoptotic path, or deactivates the competing NFκB and necrotic paths.

Table 5.1.: Minimal model extensions by one additional input.

modified update rule; $x \Uparrow (\Downarrow)$ means that B_x was replaced by $B_x \vee u_a$ ($B_x \wedge \neg u_a$)	guaranteed convergence to $\mathcal{X}_a$
RIP1 $\Downarrow$	yes
NFκB $\Downarrow$	no
CASP8 $\Uparrow$	yes
ROS $\Downarrow$	no
ATP $\Uparrow$	no
MPT $\Downarrow$	no
MOMP $\Uparrow$	yes
cIAP $\Downarrow$	no
CASP3 $\Uparrow$	yes

Partial observation

As seen in the last section, setting the additional input u_a alone is sufficient in each of the four cases in order to guarantee convergence to the apoptotic attractor. Therefore, the minimal possible set $\mathcal{X}^o$ is the empty set. If the new input u_a is however applied in combination with the old inputs TNF or FASL, some care has to be taken. Simulations show that, if u_a and TNF (or FASL) are all set to 1 at the initial living state, the system might also end up in the necrotic attractor. Therefore, we have also computed minimal sets $\mathcal{X}^o$, for which there exists a control strategy using all inputs. Using the algorithm for state attraction under partial observation, we found that it is sufficient to observe CASP8 in all four cases. For the example that u_a is applied to CASP8, the computed control strategy is shown in Figure 5.4. One can see that one has to wait until CASP8 goes up before TNF or FASL can be set.

5.4. Summary and discussion

In this chapter we have studied the problem of developing control strategies for Boolean models of gene regulation networks. Different from the two previous chapters, the goal was not the development of complete new methods, but to bring together existing approaches from the literature to answer questions which could not be dealt with before in the Boolean framework. As a result of the proposed methodology, we are now able to find a control strategy, whenever it exists, to steer the considered system to a desired attractor. The advantage compared to other approaches from the literature is, that we can explicitly consider limitations on the applicability of inputs and limitations on the observability of the state variables. Moreover, we obtain guarantees compared to the probabilistic results in a PBN setting. The example has demonstrated, that medium sized systems can easily be analyzed with this method, allowing to derive control strategies that are not obvious from visual inspection alone, or would otherwise require an in depth analysis of the system behavior. From the algorithmic perspective, the method for state attraction under partial application has been implemented as extension to the libFAUDES library and will be made publicly available. Thus, we can state that all goals formulated in Section 5.1.3 have been achieved.

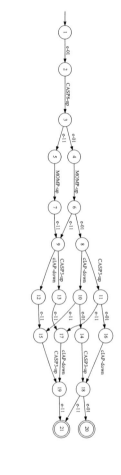

Figure 5.4.: Sequences of events guaranteeing convergence to $\mathcal{X}_a$. As TNF and FASL have the same effect, FASL has been eliminated for the purpose of this figure. The input is thus $u = (\text{TNF}, u_a)$, and with u_a upregulating CASP8. Above: case of full observation. Below: case of partial observation and $X^0 = \{\text{CASP8}\}$.

6. Conclusion

6.1. Summary

In this thesis, various problems concerning the identification, analysis and control of qualitative models of gene regulation networks have been discussed. The main goal of each chapter was the development of easily applicable algorithmic solutions to these problems. Various concepts from systems and control theory and convex optimization have been applied and extended.

Chapter 3 was concerned with the problem of identifying Boolean or discrete models of gene regulation networks. To this end, update functions which are in best agreement with a set of observations have to be reconstructed from the data. The problem of finding these update functions has first been formulated as a combinatorial problem. To find computationally more attractive formulations, a representation of Boolean and discrete functions by multiaffine polynomials has been derived. Using this polynomial representation, the combinatorial problem has been transformed into a linear optimization problem. As the main theoretical contribution, it has been shown that the optimal solutions of interest are among the vertices spanning the feasible region. Therefore, this reformulation makes the identification problem accessible to tools from convex optimization theory, and the optimal solutions can be found efficiently using a simplex solver. Compared to existing results, two essential improvements have been made. The first is the extension of the polynomial representation to discrete functions, the second is the complexity reduction achieved by the reformulation of the identification problem as a linear program. Furthermore, it has been shown that restrictions of the search space to the biologically relevant classes of unate and canalizing functions can be achieved by adding constraints to the linear optimization problem. Thereby, the beneficial property that the optimal solutions of interest are among the vertices spanning the feasible region is not lost.

While the polynomial representation for Boolean and discrete functions is most general, the number of coefficients needed to represent a given function grows exponentially in the number of arguments. Therefore, we have studied alternative and more compact representations. For the important class of unate Boolean functions, a sign-representation by affine polynomials has been suggested. While no longer all unate functions in more than three arguments can be represented by these polynomials, they have the great advantage that the number of coefficients grows only linearly with the number of arguments. Therefore, this sign-representation also leads to drastically reduced linear optimization problems compared to the general case, making the method also applicable for large systems. Furthermore, based on this sign-representation, a novel plausibility measure for the identification of robust unate Boolean update functions has been defined. It can be computed during the identification such that the search for update functions can be restricted to only the most robust ones.

In two application examples we have demonstrated that these results are not only interesting from a theoretical point of view. Several questions concerning the interaction structure of the considered example system could not be answered in reasonable time without this linear formulation.

In Chapter 4, the problems of model validations and model discrimination with respect to a desired multistable behavior were considered. Thereby, the focus on multistability as the main model property was justified with the importance of this property for numerous cellular processes. A modeling framework based on ordinary differential equations has been presented that can incorporate both, qualitative and quantitative aspects of the biological knowledge and the measurement data.

As first methodological contribution, an algorithm for the model validation problem has been presented. Based on the novel concept of Boolean decision rules, this algorithm allows to decide if a desired set of forward-invariant regions can in principle be reproduced by a hypothetical model. For a fixed maximal number of regulators, the complexity of the algorithm is linear in the system size. It is therefore a fast and efficient way to test and reject hypothetical models that are not in agreement with the desired multistability pattern.

The second contribution in this chapter is a method for model discrimination. Requiring more quantitative information about the system, the problem of verifying if a hypothetical model can in principle reproduce the desired multistability pattern, has been formulated as nonlinear feasibility problem. Moreover, a novel robustness measure for the hypothetical model has been introduced. It is based on biological robustness considerations and concepts from robust control theory, and allows to compare various model alternatives with respect to their biological plausibility. Unlike most approaches in the literature, and while based on a framework of ordinary differential equations, this measure does not depend on a given set of nominal parameters, but evaluates the interaction structure of the system itself. It can be interpreted in the following sense. The higher it is, the easier it is to find monotonic activation and inhibition functions which satisfy the model requirements. It has been shown that the computation of this measure can be formulated as optimization problem, which is even convex for two important model classes.

To demonstrate the advantage of this methodology, several hypothetical models describing one decision step in the process of cell fate determination have been analyzed. It has been found that this robustness measure correlates negatively with the number of interactions. Also, activating interactions are preferred to inhibiting ones. These findings are in good agreement with observations from other well-studied gene regulation networks.

In Chapter 5, the problem of controlling Boolean models gene regulation networks by external inputs was studied. Unlike most approaches in the literature, biologically relevant restrictions on the observable states and the applicability of inputs were addressed. We have demonstrated that the control problem under these restrictions can be described and solved in a discrete event systems framework, which was not possible in the Boolean setting. As an example of how the method can be applied, novel control strategies to guide a Boolean model of cell fate decision to a desired attractor have been derived.

In summary, this thesis contributes to the fields of identifying, analyzing and controlling models of gene regulation networks. To this end, either existing approaches have been made more efficient (Chapter 3), fundamentally new concepts and solutions have been developed (Chapter 4), or the problems have been transformed to make them applicable to existing tools (Chapter 5). The provided algorithmic solutions will be of great support for modelers at the different stages of the modeling process due to being easily applicable, generic and automatable.

6.2. Outlook

Within the context of this thesis, not all aspects relevant for the respective problem setup could be considered, and new questions were raised. Let us discuss a few future extensions which can further increase the relevance of the presented methods for applications.

Concerning the problem of finding optimal update functions for Boolean or discrete mode, large progress has been made by convexifying the related combinatorial optimization problems. In the general Boolean or discrete case however, the number of optimization variables still increases exponentially with the system size. We could reduce this complexity again for the class of unate Boolean functions by proposing sing-representations. However, for more than three variables, the generality of the sign-representation is lost. It is therefore an interesting mathematical problem with relevance to the identification of gene regulation networks to further investigate minimal polynomial representations for the class of unate Boolean functions.

Moreover, we have considered the problems of finding the individual update functions independently from each other, neglecting the global structure of the network. However, topological network properties such as connectedness, distribution of interconnections or existence of feedback loops should also be considered in this context, as often at least statistical knowledge about these properties is available. It is therefore an interesting question for future research, how such structural properties can be taken into account and treated algorithmically.

Furthermore, for the purpose of this thesis, it has been assumed that the data are available as discretized transition pairs. However, if some transitions remain unobserved, the identification problem becomes tremendously more complicated. Also, the effects of data preprocessing such as discretization should be studied in future works.

Concerning the problems of model validation and model discriminations, the focus was on the analysis of the asymptotic system behavior. However, also the transient behavior of gene regulation networks upon internal or external stimulation is of great importance. It is therefore an interesting and challenging problem for future work to extend the problem of evaluating the suitability of an interaction structure by dynamic aspects.

Finally, concerning the problem of controlling Boolean models of gene regulation networks, several extensions will increase the relevance to real biological applications. The generalization to the general discrete case is straight-forward and only requires a higher implementation effort. Model reduction techniques as by Naldi *et al.* (2011) together with the identification of conditions, under which the behavior of the reduced and the full model are equivalent, will make the presented methodology amenable to larger sized models. Furthermore, introducing biologically relevant cost functions which aim to minimize the cost of observations or control inputs, extending for example the work by Brave & Heymann (1993), will allow to study not only the questions of the existence of control strategies, but also consider optimality aspects thereof.

In conclusion, the contributions of this thesis open up new research directions. Especially, the described extensions by dynamic aspects and reduction techniques will be crucial to achieve a holistic approach toward qualitative modeling, analysis and control of large-scale, interconnected biological systems.

Appendix

A. Proofs and examples for Chapter 3

A.1. The general structure of B^I

As in the main part, assume $B^I = \{i_1, i_2, \ldots, i_m\} = i_1 \cup \bar{I}$. Then, B^I is given by

$$
B^I = \left[
\begin{array}{ccc|ccc|c|ccc}
1^1 \prod_{i \in \bar{I}} 1^1 & \cdots & 1^1 \prod_{i \in \bar{I}} 1^{d_i} & 1^2 \prod_{i \in \bar{I}} 1^1 & \cdots & 1^2 \prod_{i \in \bar{I}} 1^{d_i} & \cdots & 1^{d_{i_1}} \prod_{i \in \bar{I}} 1^1 & \cdots & 1^{d_{i_1}} \prod_{i \in \bar{I}} 1^{d_i} \\
\vdots & & \vdots & \vdots & & \vdots & & \vdots & & \vdots \\
1^1 \prod_{i \in \bar{I}} d_i^1 & \cdots & 1^1 \prod_{i \in \bar{I}} d_i^{d_i} & 1^2 \prod_{i \in \bar{I}} d_i^1 & \cdots & 1^2 \prod_{i \in \bar{I}} d_i^{d_i} & \cdots & 1^{d_{i_1}} \prod_{i \in \bar{I}} d_i^1 & \cdots & 1^{d_{i_1}} \prod_{i \in \bar{I}} d_i^{d_i} \\
\hline
2^1 \prod_{i \in \bar{I}} 1^1 & \cdots & 2^1 \prod_{i \in \bar{I}} 1^{d_i} & 2^2 \prod_{i \in \bar{I}} 1^1 & \cdots & 2^2 \prod_{i \in \bar{I}} 1^{d_i} & \cdots & 2^{d_{i_1}} \prod_{i \in \bar{I}} 1^1 & \cdots & 2^{d_{i_1}} \prod_{i \in \bar{I}} 1^{d_i} \\
\vdots & & \vdots & \vdots & & \vdots & & \vdots & & \vdots \\
2^1 \prod_{i \in \bar{I}} d_i^1 & \cdots & 2^1 \prod_{i \in \bar{I}} d_i^{d_i} & 2^2 \prod_{i \in \bar{I}} d_i^1 & \cdots & 2^2 \prod_{i \in \bar{I}} d_i^{d_i} & \cdots & 2^{d_{i_1}} \prod_{i \in \bar{I}} d_i^1 & \cdots & 2^{d_{i_1}} \prod_{i \in \bar{I}} d_i^{d_i} \\
\vdots & & \vdots & \vdots & & \vdots & & \vdots & & \vdots \\
\hline
d_{i_1}^1 \prod_{i \in \bar{I}} 1^1 & \cdots & d_{i_1}^1 \prod_{i \in \bar{I}} 1^{d_i} & d_{i_1}^2 \prod_{i \in \bar{I}} 1^1 & \cdots & d_{i_1}^2 \prod_{i \in \bar{I}} 1^{d_i} & \cdots & d_{i_1}^{d_{i_1}} \prod_{i \in \bar{I}} 1^1 & \cdots & d_{i_1}^{d_{i_1}} \prod_{i \in \bar{I}} 1^{d_i} \\
\vdots & & \vdots & \vdots & & \vdots & & \vdots & & \vdots \\
d_{i_1}^1 \prod_{i \in \bar{I}} d_i^1 & \cdots & d_{i_1}^1 \prod_{i \in \bar{I}} d_i^{d_i} & d_{i_1}^2 \prod_{i \in \bar{I}} d_i^1 & \cdots & d_{i_1}^2 \prod_{i \in \bar{I}} d_i^{d_i} & \cdots & d_{i_1}^{d_{i_1}} \prod_{i \in \bar{I}} d_i^1 & \cdots & d_{i_1}^{d_{i_1}} \prod_{i \in \bar{I}} d_i^{d_i}
\end{array}
\right].
$$

A.2. Proof of Lemma 3.8

We show that the inverse of E as in (3.13) is

$$
E_{ij}^{-1} = \begin{cases}
1 & \text{if } \Phi_{I,\chi}(j) \subseteq \Phi_{I,\chi}(i) \text{ and } |\Phi_{I,\chi}(i)| - |\Phi_{I,\chi}(j)| \text{ is even,} \\
-1 & \text{if } \Phi_{I,\chi}(j) \subseteq \Phi_{I,\chi}(i) \text{ and } |\Phi_{I,\chi}(i)| - |\Phi_{I,\chi}(j)| \text{ is odd,} \\
0, & \text{otherwise.}
\end{cases}
$$

First, as E is a lower triangular matrix, E^{-1} is a lower triangular matrix, too. Let us denote the elements of E and E^{-1} by e and i, respectively. For better readability, the indices indicating the row and the column are separated by a comma. We proceed to successively compute the elements of the i-th row of E^{-1}, starting with the diagonal element $i_{i,i}$. From $E^{-1}E = I$ we obtain

$$
i_{i,i} e_{i,i} = 1 \Rightarrow i_{i,i} = 1.
$$

To compute all elements $i_{i,l}$, $l < i$, we will use induction over $|\Phi_{I,\chi}(i)| - |\Phi_{I,\chi}(l)|$. For the base case, consider an element $i_{i,l}$ with $l < i$ and $|\Phi_{I,\chi}(i)| = |\Phi_{I,\chi}(l)|$. Note that all such elements $i_{i,l}$ are neighbors in row i. We have

$$
i_{i,l} e_{l,l} = i_{i,l} = - \sum_{l < p \leq j} i_{i,p} e_{p,l} = 0, \tag{A.1}
$$

in which the last equality follows from the construction of the matrix E in (3.13): for each $e_{p,l}$ in (A.1) we have $\Phi_{I,\chi}(p) \neq \Phi_{I,\chi}(l)$ but $|\Phi_{I,\chi}(p)| = |\Phi_{I,\chi}(l)|$. Therefore, $\Phi_{I,\chi}(l) \not\subseteq \Phi_{I,\chi}(p)$ and thus $e_{p,l} = 0$. For the inductive step assume now that for all $i_{i,l}$ with $l < i$ and $|\Phi_{I,\chi}(i)| - |\Phi_{I,\chi}(l)| < m$ it holds that

$$
\begin{aligned}
i_{i,l} &= 1 && \text{if } \Phi_{I,\chi}(l) \subseteq \Phi_{I,\chi}(i) \text{ and } |\Phi_{I,\chi}(i)| - |\Phi_{I,\chi}((l)| \text{ is even,} \\
i_{i,l} &= -1 && \text{if } \Phi_{I,\chi}(l) \subseteq \Phi_{I,\chi}(i) \text{ and } |\Phi_{I,\chi}(i)| - |\Phi_{I,\chi}((l)| \text{ is odd,} \\
i_{i,l} &= 0 && \text{otherwise,}
\end{aligned}
$$

and let us now consider an element $i_{i,l}$ with $l < i$ and $|\Phi_{I,\chi}(i)| - |\Phi_{I,\chi}(l)| = m$. Let l_q denote the largest number such that $|\Phi_{I,\chi}(i)| - |\Phi_{I,\chi}(l_q)| = q$. From the i-th row and l-th column of $E^{-1}E = I$ we obtain

$$
\begin{aligned}
i_{i,l}e_{l,l} = i_{i,l} = -&\underbrace{(i_{i,l+1}e_{l+1,l} + \ldots + i_{i,l_m}e_{l_m,l})}_{C} \\
-&\underbrace{(i_{i,l_m+1}e_{l_m+1,l} + \ldots + i_{i,l_{m-1}}e_{l_{m-1},l})}_{B_{m-1}} \\
-&\,\ldots \\
-&\underbrace{(i_{i,l_2+1}e_{l_2+1,l} + \ldots + i_{i,l_1}e_{l_1,l})}_{B_1} \\
-&\underbrace{(i_{(i,l_1+1)}e_{(l_1+1,l)} + \ldots + i_{(i,i-1)}e_{(i-1,l)})}_{A} - (i_{i,i}e_{i,l}).
\end{aligned}
\tag{A.2}
$$

Let us examine the expressions A, B_k and C closer. First, $C = 0$ as each $e_{q,l}$, $q \in \{l+1, \ldots, l_m\}$, is zero: for all such q it holds that $|\Phi_{I,\chi}(l)| - |\Phi_{I,\chi}(q)| = 0$ and $\Phi_{I,\chi}(q) \neq \Phi_{I,\chi}(l)$, which means $\Phi_{I,\chi}(l) \not\subseteq \Phi_{I,\chi}(q)$ and thus $e_{q,l} = 0$. Secondly, $A = 0$ as each $i_{i,q}$, $q \in \{l_1 + 1, \ldots, l - 1\}$ is zero: for the same reason as above we have that $\Phi_{I,\chi}(q) \not\subseteq \Phi_{I,\chi}(i)$. From the inductive step it then follows that $i_{i,q} = 0$. Finally, the expressions B_k,

$$
B_k = i_{i,l_{k-1}+1}e_{l_{k-1}+1,l} + \ldots + i_{i,l_k}e_{l_k,l}
\tag{A.3}
$$

are examined. From the inductive step it follows that all nonzero $i_{i,q}$ with $q \in \{l_{k-1} + 1, \ldots, l_k\}$ have the same sign. The same holds for all $e_{q,l}$. Therefore, let us compute $|B_k|$ first. To this end, we need to count how many summands in (A.3) are nonzero. Note that this problem is equivalent to the following combinatorial problem: given two sets $\Phi_{I,\chi}(i)$ and $\Phi_{I,\chi}(l)$ with $|\Phi_{I,\chi}(i)| - m = |\Phi_{I,\chi}(l)|$ construct all possible subsets I of $N = \{1, \ldots, n\}$ with $|\Phi_{I,\chi}(i)| - k$, $k < m$, elements and count for how many of these sets the inclusion $\Phi_{I,\chi}(l) \subseteq I \subseteq \Phi_{I,\chi}(i)$ holds. We can distinguish two cases. (i) It holds that $\Phi_{I,\chi}(l) \not\subseteq \Phi_{I,\chi}(i)$. Then this set inclusion is never true and it follows $B_k = 0$. Furthermore, $e_{i,l} = 0$ and thus $i_{i,l} = 0$, which establishes the "otherwise" case. (ii) It holds that $\Phi_{I,\chi}(l) \subset \Phi_{I,\chi}(i)$. From basic combinatorics it follows that

$$
|B_k| = \binom{m}{m-k} = \binom{m}{k}.
$$

The sign of B_k can be determined from (A.2) and we arrive at

$$B := \sum_{1 \leq k \leq m-1} B_k = -|B_1| + |B_2| - \ldots \pm |B_{m-1}|$$

$$= -\binom{m}{1} + \binom{m}{2} - \ldots \pm \binom{m}{m-1}. \tag{A.4}$$

Again, two cases have to be distinguished. (a) m is odd. Then it holds that $B = 0$ as each term $\binom{m}{y}$ of (A.4) can be canceled with a term $\binom{m}{m-y}$. It follows that $i_{i,l} = -i_{i,i}e_{i,l} = -1$ which proves this part of the lemma. (b) m is even. Using the known fact $\sum_{y=0}^{x}(-1)^y \binom{x}{y} = 0$, it follows that $B = 2$ and thus $i_{i,l} = 1$, which concludes the proof. $\square$

A.3. Proof of Lemma 3.9

Let B be a unate Boolean function and let $\mathcal{L}$ be the lattice of nodes $x \in \{0,1\}^n$ with values assigned as described in the main part. Note that, if for some node x in the lattice it holds that $B(x) = 1$, then for all nodes y, $y > x$, $B(y) = 1$ holds, too. The following two lemmas state some observations which will be useful for the proof.

Lemma A.1. *Let j be a natural number with $j < \frac{n}{2} - \frac{1}{2}$. Then, on level $j + 1$ of $\mathcal{L}$, there are more nodes x such that $B(x) = 1$ than on level j. Equivalently, let $j > \frac{n}{2} + \frac{1}{2}$. Then, on level $j - 1$ of $\mathcal{L}$, there are more nodes x such that $B(x) = 0$ than on level j.*

Proof. For each node x on level j, $0 \leq j < n$, there are $n - j$ nodes y on level $j + 1$ such that $x < y$. This means that each node in level j has $n - l$ edges to level $j + 1$. Equivalently, each node on level j, $0 < j \leq n$ has j edges to level $j - 1$.

Assume now there are k nodes x on level j, $0 < j \leq n$ with $B(x) = 1$. The smallest possible number $\tilde{k}$ of nodes y in level $j + 1$ such that $B(y) = 1$ and B is monotonic can be computed by balancing the edges starting from the k nodes x and the edges arriving at the $\tilde{k}$ nodes y:

$$\tilde{k}(j+1) = k(n-j) \tag{A.5}$$

Assuming that $\tilde{k} \leq k$, (A.5) becomes

$$k(j+1) \geq k(n-j)$$
$$j \geq \frac{n}{2} - \frac{1}{2},$$

from which the first claim follows. The second claim can be proved along the same lines. $\square$

Formulated differently, Lemma A.1 states that, if level $j + 1$ contains more nodes than level j, then there also must be more nodes x with $B(x) = 1$ on level $j + 1$ then on level j. If we consider the case that n is odd, then there are two levels, that contain the same number of nodes. This case is treated in the next lemma.

Lemma A.2. *Let n be odd. The number of nodes x in level $\hat{\imath}$ with $B(x) = 1$ and the number of nodes y in level $\hat{\imath} - 1$ with $B(y) = 1$ can only be equal if, for all nodes x and all nodes y in those levels, it holds $B(x) = B(y) = 1$. If this does not hold, the number of nodes x with $B(x) = 1$ is larger than the number of nodes y with $B(y) = 1$.*

To proof this lemma, let us first give a definition and state some preliminary observations. Two nodes x^1 and x^2 on level j of $\mathcal{L}$ are called neighbors if they agree in $j - 1$ non-zero entries. Thus, a given node x on level j has $j(n - j)$ neighbors. Next, observe that, for two given neighbors x^1 and x^2 on level j, there is exactly one node y on level $j + 1$ such that $y > x^1$ and $y > x^2$. Conversely, if x^1 and x^2 are not neighbors, there is no such node y. Also note that for a given node y on level $j + 1$, there are $\binom{j+1}{j-1}$ distinct pairs of neighboring nodes x^1 and x^2 on level j such that $y > x^1$ and $y > x^2$. Now consider two nodes, x^1 and x^k on level j, which are not neighbors. Then there is a (not necessarily unique) sequence of nodes $x^1, x^2, \ldots, x^{k-1}, x^k$ such that every two nodes x^i, x^{i+1} in that sequence are neighbors. With these preliminary observations we can proceed with the actual proof.

Proof. Assume first that there are p nodes x on level $\hat{\imath} - 1$ and $q < p$ nodes y on level $\hat{\imath}$ with $B(x) = B(y) = 1$. Collect all nodes x, $B(x) = 1$, on level $\hat{\imath} - 1$ in the set $X_1^{\hat{\imath}-1}$, and all nodes y, $B(y) = 1$, on level $\hat{\imath}$ in the set $X_1^{\hat{\imath}}$. The remaining nodes on both levels are collected in the sets $X_0^{\hat{\imath}-1}$ and $X_0^{\hat{\imath}}$, respectively. Now, note that there are $p(n - (\hat{\imath} - 1))$ edges between $X_1^{\hat{\imath}-1}$ and level $\hat{\imath}$, and $q\hat{\imath} = q(n - (\hat{\imath} - 1)) < p(n - (\hat{\imath} - 1))$ edges between the nodes in the $X_1^{\hat{\imath}}$ and the nodes on level $\hat{\imath} - 1$. This means, there must at least be one edge between two nodes $x \in X_1^{\hat{\imath}-1}$ and $y \in X_0^{\hat{\imath}}$, which is a contradiction to the monotonicity of B.

Next, assume that there are $p < \binom{n}{\hat{\imath} - 1} = \binom{n}{\hat{\imath}}$ nodes x and y on levels $\hat{\imath} - 1$ and $\hat{\imath}$, respectively, with $B(x) = B(y) = 1$. Define the same sets as before and observe that $p(n - (\hat{\imath} - 1))$ edges leave the set $X_1^{\hat{\imath}-1}$ and $p\hat{\imath} = p(n - (\hat{\imath} - 1))$ edges enter the set $X_1^{\hat{\imath}}$. Therefore, there cannot be an edge between any two nodes $x \in X_0^{\hat{\imath}-1}$ and $y \in X_1^{\hat{\imath}}$ and thus, there cannot be two nodes $x^0 \in X_0^{\hat{\imath}-1}$ and $x^1 \in X_1^{\hat{\imath}-1}$ such that there is a node $y \in X_1^{\hat{\imath}}$ with $y > x^0$ and $y > x^1$. Now, choose two arbitrary nodes $x^1 \in X_0^{\hat{\imath}-1}$ and $x^p \in X_1^{\hat{\imath}-1}$ and construct a sequence $x^1, x^2, \ldots, x^p$ of pairwise neighboring nodes. In this sequence, there must be (at least) one pair of neighbors, x^i and x^{i+1}, such that $x^i \in X_0^{\hat{\imath}-1}$ and $x^{i+1} \in X_1^{\hat{\imath}-1}$. Then, according to the preliminary observations, there must exist a unique node y on level $\hat{\imath}$ with $y > x^i$, $y > x^{i+1}$. As B is monotonic and $x^{i+1} \in X_1^{\hat{\imath}-1}$ it follows $y \in X_1^{\hat{\imath}}$, which is a contradiction to the above observation that no such state exists. $\square$

We can now continue with the proof of Lemma 3.9 and proceed in two steps. We first show that the monotonic Boolean function defined by

$$\bar{B}(x) = \begin{cases} 1 & \Phi_I(x) \geq \hat{\imath} \\ 0 & \text{otherwise} \end{cases} \qquad (A.6)$$

yields $|a_{1,\dots,1}| = \binom{n-1}{\hat{\imath}-1}$. This can be verified by calculating

$$
\begin{aligned}
a_{1,\dots,1} &= \sum_{x\in\mathcal{L}} (-1)^{H((1,\dots,1),x)} B(x) = \sum_{k=0}^{n-\hat{\imath}}(-1)^k \binom{n}{k} \\
&= \binom{n}{0} - \binom{n}{1} + \binom{n}{2} - \dots \pm \binom{n}{n-\hat{\imath}} \\
&= \binom{n-1}{0} - \left[\binom{n-1}{0} + \binom{n-1}{1}\right] + \left[\binom{n-1}{1} + \binom{n-1}{2}\right] - \qquad (A.7)\\
&\quad - \dots \pm \left[\binom{n-1}{n-\hat{\imath}-1} + \binom{n-1}{n-\hat{\imath}}\right] \\
&= \binom{n-1}{n-\hat{\imath}} = \binom{n-1}{n-1-(n-\hat{\imath})} = \pm\binom{n-1}{\hat{\imath}-1},
\end{aligned}
$$

in which the recursive formula for binomial coefficients

$$
\binom{n}{k} = \binom{n-1}{k-1} + \binom{n-1}{k}
$$

has been used. Furthermore, if n is even, (A.7) holds with "+", if n is odd, it holds with "−". It remains to show that this is indeed the maximal possible value for $|a_{1,\dots,1}|$. The two cases that n even and that n is odd will be distinguished.

(A) Let n be even. For this case, also the Boolean function defined by

$$
\hat{B}(x) = \begin{cases} 1 & \Phi_I(x) > \hat{\imath} \\ 0 & \text{otherwise} \end{cases} \qquad (A.8)
$$

will be of relevance. A similar calculation as above shows that $\bar{B}$ yields $a_{1,\dots,1} = -\binom{n-1}{\hat{\imath}-1}$, that is, the same absolute value as obtained by $\bar{B}$. Now, let B be a monotonic Boolean function $B \neq \bar{B}$, $B \neq \hat{B}$. We show how B can be successively modified until $B = \bar{B}$ or $B = \hat{B}$ while $|a_{1,\dots,1}|$ increases in each step. To begin, assume B yields $a_{1,\dots,1} \geq 0$. As $B \neq \bar{B}$, at least one of the following two cases is true: (i) there exists at least one node y, $|\Phi_I(y)| < \hat{\imath}$, $B(y) = 1$, or (ii) there exists at least one node y, $|\Phi_I(y)| \geq \hat{\imath}$, $B(y) = 0$. Let us consider case (i) first and assume $\hat{\imath} - h$ is the lowest level of $\mathcal{L}$ which contains such nodes y. If h is odd, the contribution of each such $B(y)$ to $a_{1,\dots,1}$ is negative. Thus, modifying B such that $B(y) = 0$ for each such y preserves monotonicity of B and increases $a_{1,\dots,1}$. If h is even, then the contribution of each such $B(y)$ to $a_{1,\dots,1}$ is positive. By Lemma A.1, there must however be more nodes z, $B(z) = 1$, in level $\hat{\imath} - h + 1$. Modifying B such that $B(y) = B(z) = 0$ for all such nodes y and z preserves the monotonicity of B and increases $a_{1,\dots,1}$. Using similar arguments, it can also be shown that for all nodes y described in case (ii) we can set $B(y) = 1$ while monotonicity of B is preserved and $a_{1,\dots,1}$ is increased. This procedure can be repeated until $B = \bar{B}$ is achieved. Next, assume that B yields $a_{1,\dots,1} < 0$. The above arguments need only be modified slightly to show that B can be altered successively until $B = \hat{B}$ is achieved while monotonicity of B is preserved and $a_{1,\dots,1}$ is decreased in each step. In summary, we have shown that, for even n, $\bar{B}$ and $\hat{B}$ maximize $|a_{1,\dots,1}|$.

(B) Let n be odd. In this case, also the monotonic Boolean functions $\hat{B}$ as in (A.8) and $\check{B}$ given by

$$\check{B}(x) = \begin{cases} 1 & \Phi_I(x) \geq \hat{\imath} - 1 \\ 0 & \text{otherwise} \end{cases} \tag{A.9}$$

are of relevance. One can verify by a calculation similar to (A.7) that both, $\hat{B}$ and $\check{B}$, yield $a_{1,\dots,1} = \binom{n-1}{\hat{\imath}}$, that is, a smaller absolute value than the one obtained by $\bar{B}$. Now, assume that $B \neq \bar{B}$, $B \neq \hat{B}$, $B \neq \check{B}$ yields $a_{1,\dots,1} < 0$. This case can be treated as the same case in (A) above, showing that the most negative value of $a_{1,\dots,1}$ is achieved by $\bar{B}$. Next, assume $B \neq \bar{B}$, $B \neq \hat{B}$, $B \neq \check{B}$ yields $a_{1,\dots,1} \geq 0$. We make the distinction if it holds $B(x) = 1$ for all nodes x on level $\hat{\imath} - 1$ or not. In the first case, the same arguments as before lead to the conclusion that B can successively be modified until $B = \check{B}$ is reached while $a_{1,\dots,1}$ increases in each step. In the other case, Lemma A.2 states that there are more nodes x, $B(x) = 1$ on level $\hat{\imath}$ than there are nodes y, $B(y) = 1$ on level $\hat{\imath} - 1$. Then, again using the same arguments, B can successively be modified until $B = \hat{B}$ while $a_{1,\dots,1}$ increases in each step. In summary, we have shown that the largest value for $|a_{1,\dots,1}|$ is achieved for $\bar{B}$.

B. Proofs and examples for Chapter 4

B.1. Proof of Lemma 4.15

We start with the proof of Lemma 4.15. The derivation of the inequalities according to (4.8) is straightforward and can be done in the same manner as shown in the proof of Lemma 4.14. For each rule from Table 4.2, all of the inequalities that have to be satisfied are listed in the following.

Multiplicative combinations

3. $\underline{r_i(x) = v_1(x_j) \cdot v_2(x_k)}$

     i) $B(\mathcal{I}^{x_i}) = B(\mathcal{I}^{x_j})$ and $B(\mathcal{I}^{x_k})$
$$0 < \max\{\lambda_1^{low} M_2, M_1 \lambda_2^{low}\} < \lambda_1^{high} \lambda_2^{high} < M_1 M_2$$

     ii) $B(\mathcal{I}^{x_i}) = 0$
$$0 < M_1 M_2$$

4. $\underline{r_i(x) = v_1(x_j) \cdot \mu_2(x_k)}$

     i) $B(\mathcal{I}^{x_i}) = B(\mathcal{I}^{x_j})$ and $(\text{not } B(\mathcal{I}^{x_k}))$
$$0 < \max\{\lambda_1^{low} M_2, M_1 \lambda_2^{low}\} < \lambda_1^{high} \lambda_2^{high} < M_1 M_2$$

     ii) $B(\mathcal{I}^{x_i}) = B(\mathcal{I}^{x_j})$
$$0 < \lambda_1^{low} \lambda_2^{low} < \lambda_1^{high} \lambda_2^{min} < M_1 M_2$$

     iii) $B(\mathcal{I}^{x_i}) = 0$
$$0 < M_1 M_2$$

5. $\underline{r_i(x) = \mu_1(x_j) \cdot \mu_2(x_k)}$

     i) $B(\mathcal{I}^{x_i}) = (\text{not } B(\mathcal{I}^{x_j}))$ and $(\text{not } B(\mathcal{I}^{x_k}))$
$$\lambda_1^{min} \lambda_2^{min} < \max\{M_1 \lambda_2^{low}, \lambda_1^{low} M_2\} < \lambda_1^{high} \lambda_2^{high} < M_1 M_2$$

     ii) $B(\mathcal{I}^{x_i}) = (\text{not } B(\mathcal{I}^{x_j}))$ or $(\text{not } B(\mathcal{I}^{x_k}))$
$$\lambda_1^{min} \lambda_2^{min} < \lambda_1^{low} \lambda_2^{low} < \max\{\lambda_1^{high} \lambda_2^{min}, \lambda_1^{min} \lambda_2^{high}\} < M_1 M_2$$

     iii) $B(\mathcal{I}^{x_i}) = \text{not } B(\mathcal{I}^{x_j})$
$$\lambda_1^{min} \lambda_2^{min} < \lambda_1^{high} \lambda_2^{min} < \lambda_1^{low} M_2 < M_1 M_2$$

     iv) $B(\mathcal{I}^{x_i}) = \text{not } B(\mathcal{I}^{x_k})$
$$\lambda_1^{min} \lambda_2^{min} < \lambda_1^{min} \lambda_2^{high} < M_1 \lambda_2^{low} < M_1 M_2$$

     v) $B(\mathcal{I}^{x_i}) = 1$
$$0 < \lambda_1^{min} \lambda_2^{min} < M_1 M_2$$

Additive combinations

6. $r_i(x) = v_1(x_j) + v_2(x_k)$

    i) $B(\mathcal{I}^{x_i}) = (\text{not } B(\mathcal{I}^{x_j}))$ and $(\text{not } B(\mathcal{I}^{x_k}))$

$$0 < \max\{\lambda_1^{\text{low}} + M_2, M_1 + \lambda_2^{\text{low}}\} < \lambda_1^{\text{high}} + \lambda_2^{\text{high}} < M_1 + M_2$$

    ii) $B(\mathcal{I}^{x_i}) = B(\mathcal{I}^{x_j})$ or $B(\mathcal{I}^{x_k})$

$$0 < \lambda_1^{\text{low}} + \lambda_2^{\text{low}} < \min\{\lambda_1^{\text{high}}, \lambda_2^{\text{high}}\} < M_1 + M_2$$

    iii) $B(\mathcal{I}^{x_i}) = B(\mathcal{I}^{x_j})$

$$0 < \lambda_1^{\text{low}} + M_2 < \lambda_1^{\text{high}} < M_1 + M_2$$

    iv) $B(\mathcal{I}^{x_i}) = \text{not } B(\mathcal{I}^{x_k})$

$$0 < M_1 + \lambda_2^{\text{low}} < \lambda_2^{\text{high}} < M_1 + M_2$$

    v) $B(\mathcal{I}^{x_i}) = 1$

$$0 < M_1 + M_2$$

7. $r_i(x) = v_1(x_j) + \mu_2(x_k)$

    i) $B(\mathcal{I}^{x_i}) = B(\mathcal{I}^{x_j})$ and $(\text{not } B(\mathcal{I}^{x_k}))$

$$0 < \max\{\lambda_1^{\text{low}} + M_2, M_1 + \lambda_2^{\text{low}}\} < \lambda_1^{\text{high}} + \lambda_2^{\text{high}} < M_1 + M_2$$

    ii) $B(\mathcal{I}^{x_i}) = B(\mathcal{I}^{x_j})$ or $(\text{not } B(\mathcal{I}^{x_k}))$

$$\lambda_2^{\text{min}} < \lambda_1^{\text{low}} + \lambda_2^{\text{low}} < \min\{\lambda_2^{\text{high}}, \lambda_1^{\text{high}} + \lambda_2^{\text{min}}\} < M_1 + M_2$$

    iii) $B(\mathcal{I}^{x_i}) = B(\mathcal{I}^{x_j})$

$$\lambda_2^{\text{min}} < \lambda_1^{\text{low}} + M_2 < \lambda_1^{\text{high}} + \lambda_2^{\text{min}} < M_1 + M_2$$

    iv) $B(\mathcal{I}^{x_i}) = \text{not } B(\mathcal{I}^{x_k})$

$$\lambda_2^{\text{min}} < M_1 + \lambda_2^{\text{low}} < \lambda_2^{\text{high}} < M_1 + M_2$$

    v) $B(\mathcal{I}^{x_i}) = 0$

$$0 < M_1 + M_2$$

    vi) $B(\mathcal{I}^{x_i}) = 1$

$$0 < \lambda_2^{\text{min}} < M_1 + M_2$$

8. $\underline{r_i(x) = \mu_1(x_j) + \mu_2(x_k)}$

   i) $B(\mathcal{I}^{x_i}) = (\text{not } B(\mathcal{I}^{x_j}))$ and $(\text{not } B(\mathcal{I}^{x_k}))$

   $$0 < \max\{M_1 + \lambda_2^{\text{low}}, \lambda_1^{\text{low}} + M_2\} < \lambda_1^{\text{high}} + \lambda_2^{\text{high}} < M_1 + M_2$$

   ii) $B(\mathcal{I}^{x_i}) = (\text{not } B(\mathcal{I}^{x_j}))$ or $(\text{not } B(\mathcal{I}^{x_k}))$

   $$\lambda_1^{\min} + \lambda_2^{\min} < \lambda_1^{\text{low}} + \lambda_2^{\text{low}} < \min\{\lambda_1^{\min} + \lambda_2^{\text{high}}, \lambda_1^{\text{high}} + \lambda_2^{\min}\} < M_1 + M_2$$

   iii) $B(\mathcal{I}^{x_i}) = \text{not } B(\mathcal{I}^{x_j})$

   $$\lambda_1^{\min} + \lambda_2^{\min} < \lambda_1^{\text{high}} + \lambda_2^{\min} < \lambda_1^{\text{low}} + M_2 < M_1 + M_2$$

   iv) $B(\mathcal{I}^{x_i}) = \text{not } B(\mathcal{I}^{x_k})$

   $$\lambda_1^{\min} + \lambda_2^{\min} < \lambda_1^{\min} + \lambda_2^{\text{high}} < M_1 + \lambda_2^{\text{low}} < M_1 + M_2$$

   v) $B(\mathcal{I}^{x_i}) = 1$

   $$0 < \lambda_1^{\min} + \lambda_2^{\min} < M_1 + M_2$$

Lemma 4.15 can now be proved by the following two observations. All of the above inequalities can be satisfied by an appropriate choice of critical values with the desired ordering. Moreover, for each case no other Boolean equation of the form $B(\mathcal{I}^{x_i}) = \diamond B(\mathcal{I}^{x_j}) \dagger \diamond B(\mathcal{I}^{x_k})$ can be a decision rules as this would either require that some activation function $v_a(x_b)$ is nonzero for $x_b = 0$, or that at least one of the involved monotonic functions has monotonicity properties which contradict its definition.

B.2. Proof of Lemma 4.16

As in the main part, let $\varphi_1(x)$ be a generalized activation or inhibition function, and let $B(\mathcal{I}^{x_i}) = B_1(\mathcal{I}^x)$ be a decision rule for the equation $\dot{x}_i = -\gamma_i x_i + \varphi_1(x)$. In this, $B_1(\mathcal{I}^x)$ is a Boolean expression assigning well-defined generalized Boolean intervals the values 0 or 1. Therefore, as $B(\mathcal{I}^{x_i}) = B_1(\mathcal{I}^x)$ is a decision rule, whenever $\mathcal{I}^x$ is such that $B_1(\mathcal{I}^x) = 0$, then it must hold that $\forall x \in \mathcal{I}^x : \varphi_1(x) \in [0, \lambda_{\varphi_1}^{\text{low}}]$. Also, whenever $\mathcal{I}^x$ is such that $B_1(\mathcal{I}^x) = 1$, it must hold that $\forall x \in \mathcal{I}^x : \varphi_1(x) \in [\lambda_{\varphi_1}^{\text{high}}, M_{\varphi_1}]$. Moreover, $0 < \lambda_{\varphi_1}^{\text{low}} < \lambda_{\varphi_1}^{\text{high}} < M_{\varphi_1}$.

Next, treat φ_1 as activation function if $\varphi_1(0) = 0$, and as inhibition function otherwise, and let φ_2 be an individual activation or inhibition function. Then, verifying if $B(\mathcal{I}^{x_i}) = B_1(\mathcal{I}^x) \dagger \diamond B(\mathcal{I}^{x_k})$ is a decision rule for $\dot{x}_i = -\gamma_i x_i + \varphi_1(x) \circ \varphi_2(x_k)$ leads to the same inequalities as verifying if $B(\mathcal{I}^{x_i}) = \diamond B(\mathcal{I}^{x_j}) \dagger \diamond B(\mathcal{I}^{x_k})$ is a decision rule for $\dot{x}_i = -\gamma_i x_i + v_1(x_j) \circ \varphi_2(x_k)$ (or $\dot{x}_i = -\gamma_i x_i + \mu_1(x_j) \circ \varphi_2(x_k)$, respectively). Noting that all inequalities in Section B.1 can still be satisfied if the set of critical values of one of the two monotonic functions are fixed, and only the critical values of the respective other function can be chosen freely then proves this Lemma.

B.3. A counterexample for general functions r_i

For this counterexample, consider the production term

$$r_i(x) = \mu_{i,1}(x_{j_1}) \cdot \mu_{i,2}(x_{j_2}) + \mu_{i,3}(x_{j_3}) \cdot \mu_{i,4}(x_{j_4}).$$

One possible decision rule for $r_{i,a}(x) = \mu_{i,1}(x_{j_1}) \cdot \mu_{i,2}(x_{j_2})$ is

$$B(\mathcal{I}^{x_i}) = (\text{not } B(\mathcal{I}^{x_{j_1}})) \text{ or } (\text{not } B(\mathcal{I}^{x_{j_2}})),$$

while one possible decision rule for $r_{i,b}(x) = \mu_{i,3}(x_{j_3}) \cdot \mu_{i,4}(x_{j_4})$ is

$$B(\mathcal{I}^{x_i}) = (\text{not } B(\mathcal{I}^{x_{j_3}})) \text{ or } (\text{not } B(\mathcal{I}^{x_{j_4}})).$$

Treating $r_{i,a}$ and $r_{i,b}$ as inhibition functions, one possible potential decision rule for $r_i = r_{i,a} + r_{i,b}$ in Section 4.3.2 is

$$B(\mathcal{I}^{x_i}) = [(\text{not } B(\mathcal{I}^{x_{j_1}})) \text{ or } (\text{not } B(\mathcal{I}^{x_{j_2}}))] \text{ and } [(\text{not } B(\mathcal{I}^{x_{j_3}})) \text{ or } (\text{not } B(\mathcal{I}^{x_{j_4}}))].$$

If this Boolean expression indeed was a decision rule, the inequalities

$$0 < \lambda_1^{\min}\lambda_2^{\min} < l_a := \lambda_1^{\text{low}}\lambda_2^{\text{low}}$$
$$< h_a := \min\{\lambda_1^{\min}\lambda_2^{\text{high}}, \lambda_1^{\text{high}}\lambda_2^{\min}\} < M_a := M_1 M_2, \quad \text{(B.1)}$$

$$0 < \lambda_3^{\min}\lambda_4^{\min} < l_b := \lambda_3^{\text{low}}\lambda_4^{\text{low}}$$
$$< h_b := \min\{\lambda_3^{\min}\lambda_4^{\text{high}}, \lambda_3^{\text{high}}\lambda_4^{\min}\} < M_b := M_3 M_4, \quad \text{(B.2)}$$

and

$$0 < \max\{M_a + l_b, M_b + l_a\} < h_a + h_b < M_a + M_b \quad \text{(B.3)}$$

need to be satisfied. From (B.3), it follows

$$f_a := M_a - h_a < e_b := h_b - l_b \quad \text{(B.4)}$$
$$f_b := M_b - h_b < e_a := h_a - l_a. \quad \text{(B.5)}$$

With (B.1) we can give the following approximation for f_a and e_a.

$$f_a = M_1 M_2 - \lambda_1^{\text{high}}\lambda_2^{\text{high}} > M_1 M_2 - \lambda_2^{\min}\lambda_1^{\text{high}} > M_1 M_2 - M_1\lambda_2^{\min} = M_1(M_2 - \lambda_2^{\min})$$
$$e_a = \lambda_1^{\min}\lambda_2^{\text{high}} - \lambda_1^{\text{low}}\lambda_2^{\text{low}} < \lambda_1^{\min}\lambda_2^{\text{high}} - \lambda_1^{\min}\lambda_2^{\min} = \lambda_1^{\min}(\lambda_2^{\text{high}} - \lambda_2^{\min}).$$

As $M_1 > a_1$ and $M_2 - a_2 > \lambda_2^{\text{high}} - a_2$, we have $f_a > e_a$ and equivalently $f_b > e_b$. Then, note that from $f_a < e_b$, $e_a < f_a$, and $e_b < f_b$ it follows that $e_a < f_b$, which is a contradiction to (B.4). $\square$

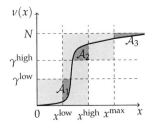

Figure 1.: Illustration of the areas $\mathcal{A}_1$, $\mathcal{A}_2$ and $\mathcal{A}_3$.

B.4. Proof of Proposition 4.23

Given a tube $T_\mathcal{N}$, an activation function $\varphi \vDash T_\mathcal{N}$ and a perturbed activation function $\nu^p \nvDash T_\mathcal{N}$. Then ν^p violates at least one of the Inequalities from Definition 4.20 and we can give the following estimates on $\|\nu - \nu^p\|_1$.

Assume that the first inequality, $\forall x \leq x^{\text{low}} : \nu(x) \leq \tau^{\text{low}}$, is violated and let $\hat{x}$ be the smallest value such that $\nu(\hat{x}) = \tau^{\text{low}}$. Then it holds that $\|\nu - \nu^p\|_1 \geq \int_{x^{\text{low}}}^{\hat{x}} (\tau^{\text{low}} - \nu(x))\,\mathrm{d}x = \mathcal{A}_1(\nu)$. Now assume that the second inequality, $\forall x \geq x^{\text{high}} : \nu(x) \geq \tau^{\text{high}}$, is violated. Let $\hat{x}$ be the smallest value such that $\nu(\hat{x}) = \tau^{\text{high}}$. Then $\|\nu - \nu^p\|_1 \geq \int_{\hat{x}}^{x^{\text{high}}} (\nu(x) - \tau^{\text{high}})\,\mathrm{d}x = \mathcal{A}_2(\nu)$. Finally, assume that the last inequality, $\forall x \leq x^{\text{max}} : \nu(x) \leq \tau^{\text{max}}$, is violated. Then, with $\hat{x}$ being the largest value such that $\nu(\hat{x}) = \tau^{\text{max}}$ it holds that $\|\nu - \nu^p\|_1 \geq \int_{x^{\text{max}}}^{\hat{x}} (\tau^{\text{max}} - \nu(x))\,\mathrm{d}x = \mathcal{A}_3(\nu)$. For an illustration of these areas, see Figure 1.

Thus, for every $\nu^p \in \mathcal{N}$ such that $\nu^p \nvDash T$ it holds that $\|\nu - \nu^p\|_1 \geq \mathcal{A}(\nu) = \min\{\mathcal{A}_1(\nu), \mathcal{A}_2(\nu), \mathcal{A}_3(\nu)\}$ and thus $\mathcal{R}^{\text{min}}(\nu, T) = \mathcal{A}(\nu)$. In order to compute $\mathcal{R}^{\text{max}}(T)$, we first derive the function $\nu^\star$ which maximizes $\mathcal{A}$.

To do so, first note that for every given $\nu \vDash T_\mathcal{N}$ it is possible to find another $\bar{\nu} \vDash T_\mathcal{N}$ which is identical with ν in $[0, \tilde{x}]$, with $\nu(\tilde{x}) = \tau^{\text{max}} - \epsilon$, ϵ arbitrarily small, such that $\mathcal{A}_3(\bar{\nu}) = \max\{\mathcal{A}_1(\bar{\nu}), \mathcal{A}_2(\bar{\nu}), \mathcal{A}_3(\bar{\nu})\}$. One possibility to achieve this is keeping $\bar{\nu}$ constant at $\tau^{\text{max}} - \epsilon$ in the interval $[\tilde{x}, \hat{x}]$ and choose $\hat{x}$ large enough. But this means that for computing the function $\nu^\star$ which maximizes $\mathcal{A}$ it is sufficient to compute the function $\nu^\star$ which maximizes $\min\{\mathcal{A}_1, \mathcal{A}_2\}$, and satisfies the modified tube $\bar{T}_\mathcal{N} = \{(x^{\text{low}}, \tau^{\text{low}}), (x^{\text{high}}, \tau^{\text{high}}), (x^{\text{max}}, \bar{\tau}^{\text{max}})\}$, with $\bar{\tau}^{\text{max}} = \tau^{\text{max}} - \epsilon$.

Now, to compute this $\nu^\star$, first define a step function

$$h_{x_s}(x) = \begin{cases} 0 & x \leq x_s \\ \bar{\tau}^{\text{max}} & x > x_s. \end{cases}$$

Next, for the given function $\bar{\nu}$, let $\tilde{x}_s$ be the smallest value such that $\bar{\nu}(\tilde{x}_s) = \frac{(\tau^{\text{high}} + \tau^{\text{low}})}{2}$. Then, $h_{\tilde{x}_s} \vDash \bar{T}_\mathcal{N}$ and with the above definitions of $\mathcal{A}_1$ and $\mathcal{A}_2$ it follows that $\min\{\mathcal{A}_1(\bar{\nu}), \mathcal{A}_2(\bar{\nu})\} \leq \min\{\mathcal{A}_1(h_{\tilde{x}_s}), \mathcal{A}_2(h_{\tilde{x}_s})\}$. This means we can always find a step function with a larger value $\mathcal{A}$ than the function $\bar{\nu}$. Next, among all possible step functions $h_{x_s} \vDash \bar{T}_\mathcal{N}$, the step function $h_{x_s^\star}$ which maximizes $\min\{\mathcal{A}_1, \mathcal{A}_2\}$ has to satisfy $\mathcal{A}_1 = \mathcal{A}_2$, i.e., $(\bar{\tau}^{\text{max}} - \tau^{\text{high}}) \cdot (x^{\text{high}} - x^{s^\star}) = \tau^{\text{low}} \cdot (x_s^\star - x^{\text{low}})$, from which it follows that $x_s^\star =$

$\frac{\tau^{\text{low}} \cdot x^{\text{low}} + (\tau^{\text{max}} - \tau^{\text{high}}) \cdot x^{\text{high}}}{\tau^{\text{low}} + (\tau^{\text{max}} - \tau^{\text{high}})}$. With this and as ϵ can be arbitrarily small we get

$$\mathcal{R}^{\text{max}}(T_{\mathcal{N}}) = \frac{\tau^{\text{low}} \cdot (\tau^{\text{max}} - \tau^{\text{high}})}{\tau^{\text{low}} + (\tau^{\text{max}} - \tau^{\text{high}})} \cdot (x^{\text{high}} - x^{\text{low}}).$$

$\square$

C. State attraction of discrete event systems

In the main part we have transferred the Boolean control problem into a state attraction problem for discrete event systems. Solutions to this problem exist for the case of full event observation (Kumar *et al.*, 1993) as well as for the case of partial observation (Schmidt & Breindl, 2014). For completeness, a summary of Schmidt & Breindl (2014) and of its main concepts and results leading to an algorithmic solution is presented in the following sections. For details, the reader is referred to this reference.

C.1. State attraction under full observation

Formally, the state attraction problem for a discrete event systems $G = (Z, \Sigma, \delta, Z_0)$ is to find a supervisor which guarantees that a predefined subset of Z' is reached in a limited number of transitions from every initial state in Z_0, and that the system remains in Z' thereafter. As the set of marked states Z_m is not relevant for the following presentation, it is omitted from the definition of G. From Brave & Heymann (1990) and Schmidt & Breindl (2014), we next recall some definitions and a solution algorithm.

First, a subset $Z' \subseteq Z$ is called an invariant set, if no transitions from a state in Z' leave this set. Based on this, the notion of a strong attractor is introduced.

Definition C.1. *Let A, Z', $A \subseteq Z' \subset Z$, be two invariant sets in G. A is called a strong attractor for Z' in G, if*

  1. *the restriction of G to the state set $Z' \backslash A$ is acyclic,*

  2. *form each state $z \in Z'$, A can be reached by a finite number of transitions.*

Next, given a set of uncontrollable events $\Sigma_u \subseteq \Sigma$, a state-feedback supervisor for G is a sub-automaton of G, from which only controllable events may be removed. The behavior of G under supervision of S can then be described as the parallel composition $G||S$ as introduced in Section 5.1.4. This means, that only controllable states in G are disabled, and that the decision about which events are to be disabled at a state z only depends on the current state z, but not on past states.

Definition C.2. *Let A, Z' be sets as defined before, and let $\Sigma_u \subseteq \Sigma$ be the set of uncontrollable events. Then, A is called a weak attractor for Z' in G, if there exists a state-feedback supervisor S for G and Σ_u, such that A is a strong attractor for Z' in S.*

Thus, A is called a weak attractor if it can be made a strong attractor by only disabling uncontrollable events in G. Given a set $A \subset Z$, there is a supremal subset of Z, denoted by $\Omega_G(Z)$, such that A is a weak attractor for $\Omega_G(A)$ in G. Algorithm C.1 computes a state-feedback supervisor $S = (Q, \Sigma, \nu, Q_0)$ for G, with $Q = \Omega_G(A)$, such that A is a strong attractor for $\Omega_G(A)$ in S. Formulated differently, for G under supervision of S it is guaranteed that the state converges to A in a finite number of transitions from every initial state in $\Omega_G(A)$, and remains there indefinitely.

C.2. State attraction under partial observation

The algorithm from the last section supposes that all events can be observed by a supervisor, that is, the supervisor always knows which state the system G is currently

---

Algorithm C.1: Solution algorithm for the state attraction problem under full observation.

Input: $G = (Z, \Sigma, \delta, Z_0)$
Output: $S = (Q, \Sigma, \nu, q_0)$

1 initialize $\Omega_{-1} = \varnothing$, $\Omega_0 = A$, $i = 0$, $Q = A$,
   $\nu(z, \sigma) = \delta(z, \sigma)$ for all states $z \in A$ and events σ that can occur at z
2 compute $\Omega_{i+1} := \Omega_i \cup \{ z \in Z \backslash \Omega_i \mid \exists \sigma \in \Sigma$
   such that $\delta(z, \sigma) \in \Omega_i \backslash \Omega_{i-1}$ and $\nexists \sigma \in \Sigma_u$ such that $\delta(z, \sigma) \notin \Omega_i \}$
3 set $Q = \Omega_{i+1}$. For each $z \in \Omega_{i+1} \backslash \Omega_i$ and each event σ with $\delta(z, \sigma) \in \Omega_i$
   set $\nu(z, \sigma) = \delta(z, \sigma)$
4 **if** $\Omega_{i+1} = \Omega_i$ **then**
5     $\lfloor$ **return** S
6 **else**
7     $\lfloor$ set $i := i + 1$ and continue with step 2

---

in. If not all events can be observed, the state attraction problem becomes more complicated. A solution to the problem of state attraction under partial observation has been presented by Schmidt & Breindl (2014), from which the most fundamental concepts and results are recalled next.

The state attraction problem is the same as before, a supervisor can however only base its decisions upon the set of observable events $\Sigma_o \subseteq \Sigma$. With the natural projection $p : \Sigma^* \rightarrow \Sigma_o^*$ as defined in Section 5.1.4, such a supervisor has to allow the same control actions after strings with the same observation, referred to as supervisor under partial observation (SPO).

The observed behavior of G, if only the events in Σ_o are observable, can be described by the deterministic observer automaton as defined next (Schmidt & Breindl, 2014; Cassandras & Lafortune, 2008).

Definition C.3. *Let $G = (Z, \Sigma, \delta, Z_m, Z_0)$ and Σ_o as defined above. The unobservable reach of a state set $Z' \subseteq Z$ and an event $\sigma \in \Sigma_o \cup \{\epsilon\}$ is*

$$UR_G(Z', \sigma) = \{ z \in Z \mid \exists z' \in Z' \text{ and } u, u' \in (\Sigma \backslash \Sigma_o)^* \text{ s.t. } z = \delta(z', u\sigma u') \}.$$

Then, the observer automaton $Obs(G) = (Obs(Z), \Sigma_o, Obs(\delta), Obs(Z_0))$ of G is defined as

- $Obs(Z_0) = UR_G(Z_0, \epsilon)$,

- $Obs(Z) \subseteq 2^Z$,

- *for each $q \in Obs(Z)$ and $\sigma \in \Sigma_o$, $Obs(\delta, \sigma) = UR_G(q, \sigma)$,*

- $q \in Obs(Z)$ *only if it is reachable from $Obs(Z_0)$.*

In analogy to the case of full observations, the notion of a weak attractor under partial observation (WAPO) is then defined. To this end, let G_z denote the automaton which is identical to G except for the set Z_0 being replaced by the single state $z \in Z_0$.

Definition C.4. *Let $G = (Z, \Sigma, \delta, z_0)$, and Σ_u and Σ_o be the sets of uncontrollable and observable events. Assume that $A \subseteq Z$ is an invariant set in G. A is called a weak attractor* *under partial observation (WAPO) for Z_0 in G with Σ_u and Σ_o if there is an automaton $S = (Q, \Sigma, \nu, q_0)$ and an integer $N \in \mathbb{N}$ such that for each $z \in Z_0$*

1. *S is an SPO for G_z with Σ_u and Σ_o,*

2. *$\forall s \in L(G_z||S)$, there is a $u \in \Sigma^\star$ such that $su \in L(G_z||S)$ and $\delta(z, su) \in A$,*

3. *$\forall s \in L(G_z||S)$, it holds that $|s| > N \Rightarrow \delta(z, s) \in A$.*

The main result by Schmidt & Breindl (2014) is then that such a SPO can always be found as subautomaton of $Obs(G)$. A few technical assumptions are introduced, most importantly that G contains no loops of unobservable events, that the set of initial states Z_0 is closed under its unobservable reach, and that all unobservable events are also uncontrollable.

Theorem C.5. *Given G, Σ_u, Σ_o, $A \subseteq Z$ as defined above. Let $Obs(G)$ be the observer automaton for G. Define the set A_0 as*

$$A_0 = \{o' \in Obs(Z) \mid \forall z \in o', \exists u \in (\Sigma \backslash \Sigma_o)^\star \text{ such that } \delta(z, u) \in A$$
$$\text{and } \forall \sigma \in \Sigma_u \cap \Sigma_o, \text{ if } \delta(z, \sigma) \text{ is defined, then } \delta(z, \sigma) \in A\}.$$

Then A is WAPO for Z_0 in G with Σ_u and Σ_o, if there exists a subautomaton $S = (Q, \Sigma_o, \nu, Z_0)$ of $Obs(G)$ such that

1. *$L(S)$ is controllable for $Obs(G)$ with $\Sigma_u \cap \Sigma_o$,*

2. *A_0 is a strong attractor for Q in S,*

3. *S is locally connected for G, Σ_0 and A.*

In this, the first condition states that S may never prevent an uncontrollable event from occurring in $Obs(G)$. For details on the third condition, we refer to Schmidt & Breindl (2014). Finally, an algorithm is presented which computes a supervisor S as in the last theorem. A supervisor for the state attraction problem under partial observation exists if $OBS(Z_0)$ is a state of the resulting automaton S.

Algorithm C.2: Solution algorithm for the state attraction problem under partial observation.

Input: $G = (Z, \Sigma, \delta, Z_0)$, $A \subseteq Z$, Σ_u, Σ_o, $o \subseteq Z$
Output: $S = (Q, \Sigma_o, \nu, o)$

1 compute $Obs(G)$ for Z_0 and Σ_o
2 compute A_o
3 initialize $\Omega_{-1} = \emptyset$, $\Omega_0 = A_o$, $i = 0$, $Q = A_o$, $\forall q \in A_o$ and $\sigma \in \Sigma_o$ such that $Obs(\delta)(q, \sigma)$ is defined, $\nu(q, \sigma) = Obs(\delta)(q, \sigma)$ if $\nexists x' \in q \setminus A$ such that $\delta(x, \sigma)$ is defined
4 compute

$$O_i := \{q \in Obs(Z) \setminus O_{i-1} | \exists \sigma \in \Sigma_o \text{ such that } Obs(\delta)(q, \sigma) \in \Omega_i \setminus \Omega_{i-1} \text{ and } \\ \nexists \sigma \in \Sigma_u \cap \Sigma_o \text{ such that } Obs(\delta)(q, \sigma) \notin \Omega_i\}$$

5 compute

$$\Omega_{i+1} := \Omega_i \cup \{q \in O_i | \forall x' \in q, \text{ either } \exists u' \in (\Sigma \setminus \Sigma_o)^\star \text{ such that } \\ \delta(x', u') \in A \text{ or } \exists u' \in (\Sigma \setminus \Sigma_o)^\star \text{ and } \sigma \in \Sigma_o \text{ such that } \\ \delta(x', u'\sigma) \text{ exists and } Obs(\delta)(q, \sigma) \in \Omega_i\}$$

6 $Q = \Omega_{i+1}$ and for all $q \in \Omega_{i+1} \setminus \Omega_i$ and $\sigma \in \Sigma_o$, $\nu(q, \sigma) = Obs(\delta)(q, \sigma)$ if $Obs(\delta)(q, \sigma) \in \Omega_i$
7 **if** $\Omega_{i+1} = \Omega_i$ **then**
8   **return** S
9 **else**
10   set $i := i + 1$ and continue with step 4

Bibliography

Akutsu, T., Hayashida, M., Ching, W.-K., & Ng, M. K. (2007). Control of Boolean networks: hardness results and algorithms for tree structured networks. *Journal of Theoretical Biology*, 244(4), 670 – 679.

Akutsu, T., Miyano, S., & Kuhara, S. (1999). Identification of genetic networks from a small number of gene expression patterns under the Boolean network model. In *Proceedings of the Pacific Symposium on Biocomputing*. vol. 4, 17–28.

Akutsu, T., Miyano, S., & Kuhara, S. (2000). Inferring qualitative relations in genetic networks and metabolic pathways. *Bioinformatics*, 16(8), 727–734.

Albert, R. & Othmer, H. G. (2003). The topology of the regulatory interactions predicts the expression pattern of the segment polarity genes in drosophila melanogaster. *Journal of Theoretical Biology*, 223(1), 1 – 18.

Aldana, M., Balleza, E., Kauffman, S., & Resendiz, O. (2007). Robustness and evolvability in genetic regulatory networks. *Journal of Theoretical Biology*, 245(3), 433–448.

Angeli, D. & Sontag, E. D. (2003). Monotone control systems. *IEEE Transactions on Automatic Control*, 48, 1684–1698.

Angeli, D. & Sontag, E. D. (2004). Multi-stability in monotone input/ouptut systems. *Systems & Control Letters*, 51, 185–202.

Ay, A. & Arnosti, D. N. (2011). Mathematical modeling of gene expression: a guide for the perplexed biologist. *Critical Reviews in Biochemistry and Molecular Biology*, 46(2), 137–151.

Baldissera, F. L. & Cury, J. E. R. (2012). Application of supervisory control theory to guide cellular dynamics. In *Proceedings of the Workshop on Discrete Event Systems (WODES)*. 384–389.

Balleza, E., Alvarez-Buylla, E. R., Chaos, A., Kauffman, S., Shmulevich, I., & Aldana, M. (2008). Critical dynamics in genetic regulatory networks: examples from four kingdoms. *PLoS One*, 3(6), e2456.

Batt, G., de Jong, H., Page, M., & Geiselmann, J. (2008). Symbolic reachability analysis of genetic regulatory networks using discrete abstractions. *Automatica*, 44(4), 982–989.

Beard, D. A., Liang, S., & Qian, H. (2002). Energy balance for analysis of complex metabolic networks. *Biophysical Journal*, 83(1), 79–86.

Ben D. MacArthur, A. M. & Lemischka, I. R. (2009). Systems biology of stem cell fate and cellular reprogramming. *Nature Reviews Molecular Cell Biology*, 10, 672–681.

Blanchini, F. (1999). Set invariance in control. *Automatica*, 35, 1747–1767.

Blanchini, F. & Franco, E. (2011). Structurally robust biological networks. *BMC Systems Biology*, 5(1), 74.

Bornholdt, S. (2008). Boolean network models of cellular regulation: prospects and limitations. *Journal of the Royal Society Interface*, 5, S85–S94.

Boyd, S. & Vandenberghe, L. (2004). *Convex Optimization.* Cambridge: Cambridge University Press.

Brave, Y. & Heymann, M. (1990). Stabilization of discrete-event processes. *International Journal of Control, 51*(5), 1101–1117.

Brave, Y. & Heymann, M. (1993). On optimal attraction of discrete-event processes. *Information Sci, 67*(3), 245–276.

Breindl, C. & Allgöwer, F. (2009). Verification of multistability in gene regulation networks: A combinatorial approach. In *Proceedings of the 48th IEEE Conference on Decision and Control (CDC).* 5637–5642.

Breindl, C., Chaves, M., & Allgöwer, F. (2013). A linear reformulation of Boolean optimization problems and its application to the problem of estimating the structure of gene regulation networks. In *Proceedings of the 52nd IEEE Conference on Decision and Control (CDC).* 733 – 738.

Breindl, C., Chaves, M., Gouzé, J.-L., & Allgöwer, F. (2012). Structure estimation for unate Boolean models of gene regulation networks. In *Proceedings of the 16th IFAC Symposium on System Identification (SYSID).* 1725–1730.

Breindl, C., Schittler, D., Waldherr, S., & Allgöwer, F. (2011a). Structural requirements and discrimination of cell differentiation networks. In *Proceedings of the 18th IFAC World Congress.* 11767–11772.

Breindl, C., Waldherr, S., & Allgöwer, F. (2010). A robustness measure for the stationary behavior of qualitative gene regulation networks. In *Proceedings of the 11th symposium on computer applications in biotechnology (CAB).* 36 – 41.

Breindl, C., Waldherr, S., Hausser, A., & Allgöwer, F. (2009). Modeling cofilin mediated regulation of cell migration as a biochemical two-input switch. In *Proceedings of the 3rd Foundations of Systems Biology in Engineering Conference (FOSBE).* 60–63.

Breindl, C., Waldherr, S., Wittmann, D. M., Theis, F. J., & Allgöwer, F. (2011b). Steady-state robustness of qualitative gene regulation networks. *International Journal of Robust and Nonlinear Control, 21*(15), 1742–1758.

Calzone, L., Tournier, L., Fourquet, S., Thieffry, D., Zhivotovsky, B., Barillot, E., & Zinovyev, A. (2010). Mathematical modelling of cell-fate decision in response to death receptor engagement. *PLoS Computational Biology, 6*(3), e1000702.

Candès, E. J., Romberg, J. K., & Tao, T. (2006). Stable signal recovery from incomplete and inaccurate measurements. *Communications on Pure and Applied Mathematics, 59*(8), 1207–1223.

Candès, E. J., Wakin, M. B., & Boyd, S. P. (2008). Enhancing sparsity by reweighted l_1 minimization. *Journal of Fourier Analysis and Applications, 14*(5), 877–905.

Casey, R., de Jong, H., & Gouzé, J.-L. (2006). Piecewise-linear models of genetic regulatory networks: Equilibria and their stability. *Journal of Mathematical Biology, 52*(1), 27–56.

Cassandras, C. G. & Lafortune, S. (2008). *Introduction to discrete event systems.* New York: Springer.

Chaouiya, C. & Remy, É.. (2013). Logical modelling of regulatory networks, methods and applications. *Bulletin of Mathematical Biology, 75*(6), 891–895.

Chaves, M., Albert, R., & Sontag, E. D. (2005). Robustness and fragility of Boolean models for genetic regulatory networks. *Journal of Theoretical Biology, 235*(3), 431–449.

Chaves, M., Eißing, T., & Allgöwer, F. (2009). Regulation of apoptosis via the NFκB pathway: modeling and analysis. *Dynamics on and of complex networks: applications to biology, computer science and the social sciences*, Birkhauser, Boston, Modeling and Simulation in Science, Engineering and Technology. 19–34.

Chaves, M., Eissing, T., & Allgöwer, F. (2008). Bistable biological systems: A characterization through local compact input-to-state stability. *IEEE Transactions on Automatic Control, Special Issue on Systems Biology, 53*, 87–100.

Chaves, M., Tournier, L., & Gouzé, J.-L. (2010). Comparing Boolean and piecewise affine differential models for genetic networks. *Acta Biotheoretica, 58*(2-3), 217–232.

Chen, P. C. & Weng, Y. (2009). Automaton models of computational genetic regulatory networks with combinatorial gene-protein logical interactions. In *Proceedings of the International Joint Conference on Bioinformatics, Systems Biology and Intelligent Computing*. 301–307.

Chen, P. C. & Weng, Y. (2011). Automaton models of computational genetic regulatory networks with combinatorial gene–protein interactions. *Biosystems, 106*(1), 19 – 27.

Cheng, D. & Qi, H. (2010). A linear representation of dynamics of Boolean networks. *IEEE Transactions on Automatic Control, 55*(10), 2251–2258.

Cheng, D., Qi, H., Li, Z., & Liu, J. B. (2011). Stability and stabilization of Boolean networks. *International Journal of Robust and Nonlinear Control, 21*(2), 134–156.

Cheng, D. & Zhao, Y. (2011). Identification of Boolean control networks. *Automatica, 47*(4), 702 – 710.

Cho, K.-H., Choo, S.-M., Jung, S., Kim, J.-R., Choi, H.-S., & Kim, J. (2007). Reverse engineering of gene regulatory networks. *IET Systems Biology, 1*(3), 149–163.

Cooper, N., Belta, C., & Julius, A. (2011). Genetic regulatory network identification using multivariate monotone functions. In *Decision and Control and European Control Conference (CDC-ECC), 2011 50th IEEE Conference on*. 2208 –2213.

Crick, F. (1970). Central dogma of molecular biology. *Nature, 227*, 561–563.

Datta, A., Choudhary, A., Bittner, M. L., & Dougherty, E. R. (2003). External control in markovian genetic regulatory networks. *Machine Learning, 52*, 169–191.

Datta, A., Choudhary, A., Bittner, M. L., & Dougherty, E. R. (2004). External control in markovian genetic regulatory networks: the imperfect information case. *Bioinformatics, 20*(6), 924–930.

Datta, A., Pal, R., & Dougherty, E. R. (2006). Intervention in probabilistic gene regulatory networks. *Current Bioinformatics, 1*(1), 167–184.

de Jong, H. (2002). Modeling and simulation of genetic regulatory systems: a literature review. *Journal of Computational Biology, 9*, 67–103.

de Jong, H., Gouze, J.-u., Hernandez, C., Page, M., Sari, T., & Geiselmann, J. (2004). Qualitative simulation of genetic regulatory networks using piecewise-linear models. *Bulletin of Mathematical Biology, 66*, 301–340.

Derrida, B. & Stauffer, D. (1986). Phase transitions in two-dimensional kauffman cellular automata. *EPL (Europhysics Letters), 2*(10), 739.

Donoho, D. L. (2006). Compressed sensing. *IEEE Transactions on Information Theory,*

52(4), 1289–1306.

Edwards, J. S., Ibarra, R. U., & Palsson, B. O. (2001). In silico predictions of escherichia coli metabolic capabilities are consistent with experimental data. *Nature Biotechnology*, *19*, 125–130.

Eißing, T., Allgöwer, F., & Bullinger, E. (2005). Robustness properties of apoptosis models with respect to parameter variations and intrinsic noise. *IET Systems Biology*, *152*(4), 221– 228.

Eissing, T., Conzelmann, H., Gilles, E. D., Allgöwer, F., Bullinger, E., & Scheurich, P. (2004). Bistability analyses of a caspase activation model for receptor-induced apoptosis. *The Journal of Biological Chemistry*, *279*, 36892–36897.

Faisal, S., Lichtenberg, G., Trump, S., & Attinger, S. (2010). Structural properties of continuous representations of Boolean functions for gene network modelling. *Automatica*, *46*(12), 2047 – 2052.

Faisal, S., Lichtenberg, G., & Werner, H. (2008). Polynomial models of gene dynamics. *Neurocomputing*, *71*(13-15), 2711 – 2719.

Foster, D. V., Foster, J. G., Huang, S., & Kauffman, S. A. (2009). A model of sequential branching in hierarchical cell fate determination. *Journal of Theoretical Biology*, *260*, 589–597.

Franke, D. (1994). *Sequentielle Systeme*. Braunschweig: Vieweg.

Gardner, T. S. & Faith, J. J. (2005). Reverse-engineering transcription control networks. *Physics of Life Reviews*, *2*(1), 65 – 88.

Glass, L. (1975). Classification of biological networks by their qualitative dynamics. *Journal of Theoretical Biology*, *54*(1), 85–107.

Glass, L. & Kauffman, S. A. (1973). The logical analysis of continuous, non-linear biochemical control networks. *Journal of Theoretical Biology*, *39*(1), 103–129.

Grefenstette, J., Kim, S., & Kauffman, S. (2006). An analysis of the class of gene regulatory functions implied by a biochemical model. *Biosystems*, *84*(2), 81 – 90.

Harris, S. E., Sawhill, B. K., Wuensche, A., & Kauffman, S. (2002). A model of transcriptional regulatory networks based on biases in the observed regulation rules. *Complexity*, *7*(4), 23–40.

Hecker, M., Lambeck, S., Toepfer, S., van Someren, E., & Guthke, R. (2009). Gene regulatory network inference: data integration in dynamic models—a review. *Biosystems*, *96*(1), 86 – 103.

Hickman, G. J. & Hodgman, T. C. (2009). Inference of gene regulatory networks using Boolean-network inference methods. *Journal of Bioinformatics and Computational Biology*, *07*(06), 1013–1029.

Ingolia, N. T. (2004). Topology and robustness in the drosophila segment polarity network. *PLoS Biology*, *2*(6), e123.

Jacobsen, E. W. & Cedersund, G. (2008). Structural robustness of biochemical network models-with application to the oscillatory metabolism of activated neutrophils. *IET Systems Biology*, *2*(1), 39–47.

Jarrah, A. S., Laubenbacher, R., Stigler, B., & Stillman, M. (2007a). Reverse-engineering of polynomial dynamical systems. *Advances in Applied Mathematics*, *39*(4), 477 – 489.

Jarrah, A. S., Raposa, B., & Laubenbacher, R. (2007b). Nested canalyzing, unate cascade, and polynomial functions. *Physica D: Nonlinear Phenomena, 233*(2), 167 – 174.

Kauffman, S., Peterson, C., Samuelsson, B., & Troein, C. (2003). Random Boolean network models and the yeast transcriptional network. *Proceedings of the National Academy of Sciences (PNAS), 100*(25), 14796–14799.

Kauffman, S., Peterson, C., Samuelsson, B., & Troein, C. (2004). Genetic networks with canalyzing Boolean rules are always stable. *Proceedings of the National Academy of Sciences of the United States of America (PNAS), 101*(49), 17102–17107.

Kauffman, S. A. (1969). Metabolic stability and epigenesis in randomly constructed genetic nets. *Journal of Theoretical Biology, 22,* 437–462.

Kauffman, S. A. (1993). *The origins of order: self-organization and selection in evolution.* New York: Oxford University Press.

Kholodenko, B. N., Kiyatkin, A., Bruggeman, F. J., Sontag, E., Westerhoff, H. V., & Hoek, J. B. (2002). Untangling the wires: A strategy to trace functional interactions in signaling and gene networks. *Proceedings of the National Academy of Sciences of the United States of America (PNAS), 99*(20), 12841–12846.

Kitano, H. (2004). Biological robustness. *Nature Reviews Genetics, 5,* 826–837.

Klamt, S., Saez-Rodriguez, J., Lindquist, J. A., Simeoni, L., & Gilles, E. D. (2006). A methodology for the structural and functional analysis of signaling and regulatory networks. *BMC Bioinformatics, 7,* 56.

Klemm, K. & Bornholdt, S. (2005). Topology of biological networks and reliability of information processing. *Proceedings of the National Academy of Sciences of the United States of America (PNAS), 102*(51), 18414–18419.

Klipp, E., Liebermeister, W., Wierling, C., Kowald, A., Lehrach, H., & Herwig, R. (2009). *Systems biology: a textbook.* Weinheim: Wiley-VCH.

Krumsiek, J., Marr, C., Schroeder, T., & Theis, F. J. (2011). Hierarchical differentiation of myeloid progenitors is encoded in the transcription factor network. *PLoS ONE, 6*(8), e22649.

Kumar, R., Garg, V., & Marcus, S. I. (1993). Language stability and stabilizability of discrete event dynamical systems. *SIAM Journal on Control and Optimization, 31*(5), 1294–1320.

Kwon, Y.-K. & Cho, K.-H. (2008). Quantitative analysis of robustness and fragility in biological networks based on feedback dynamics. *Bioinformatics, 24*(7), 987–994.

Leclerc, R. D. (2008). Survival of the sparsest: robust gene networks are parsimonious. *Molecular Systems Biology, 4,* 213.

Lemon, G., Waters, S. L., Rose, F. R., & Kinga, J. R. (2007). Mathematical modelling of human mesenchymal stem cell proliferation and differentiation inside artificial porous scaffolds. *Journal of Theoretical Biology, 249,* 543–553.

Li, F. & Sun, J. (2011). Stability and stabilization of multivalued logical networks. *Nonlinear Analysis: Real World Applications, 12*(6), 3701 – 3712.

Liang, S., Fuhrmann, S., & Somogyi, R. (1998). Reveal, a general reverse engineering algorithm for inference of genetic network architectures. In *Proceedings of the Pacific Symposium on Biocomputing.* vol. 3, 18–29.

Liu, Q. (2012). An optimal control approach to probabilistic Boolean networks. *Physica*

A: Statistical Mechanics and its Applications, *391*(24), 6682 – 6689.

Lodish, H. (2004). *Molecular cell biology*. New York: Freeman.

Moor, T., Schmidt, K., & Perk, S. (2008). libfaudes – an open source c++ library for discrete event systems. In *Proceedings of the Workshop on Discrete Event Systems (WODES)*. 125 – 130.

Muroga, S. (1971). *Threshold logic and its applications*. New York: Wiley-Interscience.

Naldi, A., Remy, E., Thieffry, D., & Chaouiya, C. (2011). Dynamically consistent reduction of logical regulatory graphs. *Theoretical Computer Science*, *412*(21), 2207 – 2218.

Ng, M., Zhang, S.-Q., Ching, W.-K., & Akutsu, T. (2006). A control model for markovian genetic regulatory networks. *Transactions on Computational Systems Biology V*, Berlin: Springer, vol. 4070 of *Lecture Notes in Computer Science*. 36–48.

Nikolajewa, S., Friedel, M., & Wilhelm, T. (2007). Boolean networks with biologically relevant rules show ordered behavior. *Biosystems*, *90*(1), 40 – 47.

Ogden, S. K., Jr., M. A., Stegman, M. A., & Robbins, D. J. (2004). Regulation of hedgehog signaling: a complex story. *Biochemical Pharmacology*, *67*(5), 805 – 814.

Ozbudak, E. M., Thattai, M., Lim, H. N., Shraiman, B. I., & van Oudenaarden, A. (2004). Multistability in the lactose utilization network of escherichia coli. *Nature*, *427*, 747–740.

Pal, R., Datta, A., Bittner, M. L., & Dougherty, E. R. (2005). Intervention in context-sensitive probabilistic Boolean networks. *Bioinformatics*, *21*(7), 1211–1218.

Parrilo, P. A. (2003). Semidefinite programming relaxations for semialgebraic problems. *Mathematical Programming*, *96*(2), 293–320.

Peltier, J. & Schaffer, D. (2010). Systems biology approaches to understand stem cell fate choice. *IET Systems Biology*, *4*, 1–11.

Polpitiya, A., Cobb, J., & Ghosh, B. (2005). Genetic regulatory networks and co-regulation of genes: A dynamic model based approach. *New Directions and Applications in Control Theory*, Berlin: Springer, vol. 321 of *Lecture Notes in Control*. 291–304.

Polynikis, A., Hogan, S. J., & di Bernardo, M. (2009). Comparing different ODE modelling approaches for gene regulatory networks. *Journal of Theoretical Biology*, *261*(4), 511 – 530.

Porreca, R., Cinquemani, E., Lygeros, J., & Ferrari-Trecate, G. (2010). Structural identification of unate-like genetic network models from time-lapse protein concentration measurements. In *Proceedings of the 49th IEEE Conference on Decision and Control (CDC)*. 2529–2534.

Porreca, R., Cinquemani, E., Lygeros, J., & Ferrari-Trecate, G. (2012). Invalidation of the structure of genetic network dynamics: a geometric approach. *International Journal of Robust and Nonlinear Control*, *22*(10), 1140–1156.

Prill, R. J., Iglesias, P. A., & Levchenko, A. (2005). Dynamic properties of network motifs contribute to biological network organization. *PLoS Biology*, *3*(11), e343.

Purnick, P. E. M. & Weiss, R. (2009). The second wave of synthetic biology: from modules to systems. *Nature Reviews Molecular Cell Biology*, *10*, 410–422.

Raeymaekers, L. (2002). Dynamics of Boolean networks controlled by biologically meaningful functions. *Journal of Theoretical Biology*, *18*(3), 331 – 341.

Ramadge, P. J. & Wonham, W. M. (1987). Supervisory control of a class of discrete event processes. *SIAM Journal on Control and Optimization*, *25*(1), 206–230.

Rodrigo, G. & Elena, S. F. (2011). Structural discrimination of robustness in transcriptional feedforward loops for pattern formation. *PLoS ONE*, *6*(2), e16904.

Roeder, I. & Glauche, I. (2006). Towards an understanding of lineage specification in hematopoietic stem cells: a mathematical model for the interaction of transcription factors GATA-1 and PU.1. *Journal of Theoretical Biology*, *241*(4), 852–865.

Ropers, D., de Jong, H., Page, M., Schneider, D., & Geiselmann, J. (2006). Qualitative simulation of the carbon starvation response in escherichia coli. *Biosystems*, *84*(2), 124 – 152.

S. Huang, G. M., Y.-P. Guo & Enver, T. (2007). Bifurcation dynamics in lineage-commitment in bipotent progenitor cells. *Developmental Biology*, *305*, 695–713.

Saez-Rodriguez, J., Simeoni, L., Lindquist, J. A., Hemenway, R., Bommhardt, U., Arndt, B., Haus, U.-U., Weismantel, R., Gilles, E. D., Klamt, S., & Schraven, B. (2007). A logical model provides insights into t cell receptor signaling. *PLoS Computational Biology*, *3*(8), e163.

Saks, M. E. (1993). *Surveys in Combinatorics*, Cambridge: Cambridge University Press, chap. Slicing the hypercube. 211–255.

Sánchez, L. & Thieffry, D. (2001). A logical analysis of the drosophila gap-gene system. *Journal of Theoretical Biology*, *211*(2), 115 – 141.

Schittler, D., Hasenauer, J., Allgöwer, F., & Waldherr, S. (2010). Cell differentiation modeled via a coupled two-switch regulatory network. *Chaos*, *20*(4), 045121.

Schlatter, R., Schmich, K., Avalos Vizcarra, I., Scheurich, P., Sauter, T., Borner, C., Ederer, M., Merfort, I., & Sawodny, O. (2009). ON/OFF and beyond - a Boolean model of apoptosis. *PLoS Computational Biology*, *5*(12), e1000595.

Schmidt, K. W. & Breindl, C. (2014). A framework for state attraction of discrete event systems under partial observation. *Information Sciences*, *281*(10), 265–280.

Schrijver, A. (1999). *Theory of linear and integer programming*. Chichester, England: Wiley.

Shmulevich, I., Dougherty, E. R., Kim, S., & Zhang, W. (2002a). Probabilistic Boolean networks: a rule-based uncertainty model for gene regulatory networks. *Bioinformatics*, *18*(2), 261–274.

Shmulevich, I., Dougherty, E. R., & Zhang, W. (2002b). Gene perturbation and intervention in probabilistic Boolean networks. *Bioinformatics*, *18*(10), 1319–1331.

Snoussi, E. H. (1989). Qualitative dynamics of piecewise-linear differential equations: a discrete mapping approach. *Dynamics and Stability of Systems*, *4*(3-4), 565–583.

Somogyi, R. & Sniegoski, C. A. (1996). Modeling the complexity of genetic networks: Understanding multigenic and pleiotropic regulation. *Complexity*, *1*(6), 45–63.

Steggles, L., Banks, R., & Wipat, A. (2006). Modelling and analysing genetic networks: From Boolean networks to Petri nets. *Computational Methods in Systems Biology*, Springer Berlin Heidelberg, vol. 4210 of *Lecture Notes in Computer Science*. 127–141.

Sugita, M. (1961). Functional analysis of chemical systems in vivo using a logical circuit equivalent. *Journal of Theoretical Biology, 1,* 415–30.

Sugita, M. (1963). Functional analysis of chemical systems in vivo using a logical circuit equivalent. ii. the idea of a molecular automaton. *Journal of Theoretical Biology, 4*(2), 179 – 192.

Thomas, R. (1973). Boolean formalization of genetic control circuits. *Journal of Theoretical Biology, 42*(3), 563–585.

Thomas, R. (1991). Regulatory networks seen as asynchronous automata: A logical description. *Journal of Theoretical Biology, 153*(1), 1 – 23.

Thomas, R. & D'Ari, R. (1990). *Biological Feedback.* Boca Raton: CRC Press.

Thomas, R. & Kaufman, M. (2001). Multistationarity, the basis of cell differentiation and memory. II. logical analysis of regulatory networks in terms of feedback circuits. *Chaos, 11*(1), 180–195.

Thomas, R., Thieffry, D., & Kaufman, M. (1995). Dynamical behaviour of biological regulatory networks–I. Biological role of feedback loops and practical use of the concept of the loop-characteristic state. *Bulletin of Mathematical Biology, 57*(2), 247–276.

Veliz-Cuba, A., Jarrah, A. S., & Laubenbacher, R. (2010). Polynomial algebra of discrete models in systems biology. *Bioinformatics, 26*(13), 1637–1643.

von Dassow, G., Meir, E., Munro, E. M., & Odell, G. M. (2000). The segment polarity network is a robust developmental module. *Nature, 406*(6792), 188–192.

Wajant, H., Pfizenmaier, K., & Scheurich, P. (2003). Tumor necrosis factor signaling. *Cell Death and Differentiation, 10,* 45–65.

Waldherr, S., Allgöwer, F., & Jacobsen, E. W. (2009). Kinetic perturbations as robustness analysis tool for biochemical reaction networks. In *Proceedings of the 48th IEEE Conference on Decision and Control (CDC).* 4572—4577.

Waldherr, S. & Allgöwer, F. (2011). Robust stability and instability of biochemical networks with parametric uncertainty. *Automatica, 47*(6), 1139 – 1146.

Waldherr, S., Findeisen, R., & Allgöwer, F. (2008). Global sensitivity analysis of biochemical reaction networks via semidefinite programming. In *Proceedings of the 17th IFAC World Congress.* Seoul, Korea, 9701–9706.

Waldherr, S., Hasenauer, J., Doszczak, M., Scheurich, P., & Allgöwer, F. (2011). Global uncertainty analysis for a model of TNF-induced NF-κB signalling. *Advances in the Theory of Control, Signals and Systems with Physical Modeling,* Springer Berlin / Heidelberg, vol. 407 of *Lecture Notes in Control and Information Sciences.* 365–377.

Walter, C., Parker, R., & Ycas, M. (1967). A model for binary logic in biochemical systems. *Journal of Theoretical Biology, 15*(2), 208 – 217.

Wittmann, D., Krumsiek, J., Saez-Rodriguez, J., Lauffenburger, D., Klamt, S., & Theis, F. (2009a). Transforming Boolean models to continuous models: methodology and application to t-cell receptor signaling. *BMC Systems Biology, 3*(1), 98.

Wittmann, D. M., Bloechl, F., Truembach, D., Wurst, W., Prakash, N., & Theis, F. J. (2009b). Spatial analysis of expression patterns predicts genetic interactions at the mid-hindbrain boundary. *PLoS Computational Biology, 5*(11), e1000569.

Wittmann, D. M., Marr, C., & Theis, F. J. (2010). Biologically meaningful update rules increase the critical connectivity of generalized kauffman networks. *Journal of Theoretical Biology, 266*(3), 436–448.

Xiong, W. & Ferrell Jr., J. E. (2003). A positive-feedback-based bistable memory module that governs a cell fate decision. *Nature, 426,* 460–465.

Yagil, G. & Yagil, E. (1971). On the relation between effector concentration and the rate of induced enzyme synthesis. *Biophysical Journal, 11*(1), 11 – 27.

Zhang, S.-Q., Hayashida, M., Akutsu, T., Ching, W.-K., & Ng, M. K. (2007). Algorithms for finding small attractors in Boolean networks. *EURASIP Journal on Bioinformatics and Systems Biology,* 20180.